FUNDAMENTALS OF POWER ELECTRONICS WITH MATLAB

FUNDAMENTALS OF POWER ELECTRONICS WITH MATLAB

RANDALL SHAFFER

CHARLES RIVER MEDIA
Boston, Massachusetts

Cover Design: Tyler Creative

CHARLES RIVER MEDIA
25 Thomson Place
Boston, Massachusetts 02210
617-757-7900
617-757-7969 (FAX)
crm.info@thomson.com
www.charlesriver.com

This book is printed on acid-free paper.

Randall Shaffer. *Fundamentals of Power Electronics with MATLAB.*
ISBN: 1-58450-852-3

Library of Congress Cataloging-in-Publication Data
Shaffer, Randall Alan, 1959-
  Fundamentals of power electronics with MATLAB / Randall Shaffer. --
1st ed.
     p. cm.
  Includes index.
  ISBN 1-58450-852-3 (hardcover with cd : alk. paper)
  1. Power electronics. 2. MATLAB. 3. Electric circuit analysis--Data
processing. I. Title.
  TK7881.15.S52 2007
  621.31'7--dc22
                          2006018747

Printed in the United States of America
06 7 6 5 4 3 2 First Edition

CHARLES RIVER MEDIA titles are available for site license or bulk purchase by institutions, user groups, corporations, etc. For additional information, please contact the Special Sales Department at 800-347-7707.

Requests for replacement of a defective CD-ROM must be accompanied by the original disc, your mailing address, telephone number, date of purchase and purchase price. Please state the nature of the problem, and send the information to CHARLES RIVER MEDIA, 25 Thomson Place, Boston, Massachusetts 02210. CRM's sole obligation to the purchaser is to replace the disc, based on defective materials or faulty workmanship, but not on the operation or functionality of the product.

To my son, Alexander.

Contents

Preface

The approach to the subject of this text is both different *and* new. It is different because of the use of MATLAB for circuit computations, simulations, waveform plots, and spectrum analysis. The mathematical problems of power electronics are typically numerical. Many of the equations that arise have no algebraic solution. The engineer must seek either a numerical or graphical solution. MATLAB is a natural for power electronics because the circuit equations transport readily to the scripted code of an m-file. In addition to the numerical capabilities of MATLAB, the variety of graphical features greatly assists in the understanding of a circuit problem.

Why use MATLAB instead of time-honored and capable PSpice? This is a good question that, fortunately, has some good answers. Numerical and graphical talents aside, MATLAB is an excellent pedagogical tool that bridges the gap between circuit analysis and circuit simulation. PSpice has no such quality because the circuit information must be entered either as a netlist (a list of terminal and node names) or as a drafted schematic. With MATLAB, the circuit *equations* are entered into a script file with little more alteration of the equation text than the insertion of the mathematical operators. PSpice is a clear choice for circuit simulation at the component level. If an engineer must know how a specific transistor, for example, affects the behavior of a circuit, PSpice provides the solution because the transistor model is part of the simulation. From a circuit functional point of view, however, MATLAB far excels in terms of ease of use and speed of simulation, without the infamous convergence problems so often encountered with PSpice. Admittedly, MATLAB requires programming, while PSpice does not. However, the toolbox provided on the companion CD-ROM greatly eases the programming task. Many computations and simulations are reduced to a simple function call.

The approach to the subject of power electronics in this text is also new. The periodic waveforms of the circuits are modeled as angular functions rather than as time functions. This notation provides concise mathematical elegance in that ωt is shortened to θ, and all waveform periods become 2π. The material on DC-DC converters is also new. A new body of theory for two-voltage-level DC-DC converters is developed on the introduction of the λ-ratio. The λ-ratio is the ratio of the circuit storage inductance to the inductance required for continuous inductor current operation. The ratio not only defines the boundary between the continuous and discontinuous conduction modes, but the DC-DC converter equations are also a function of λ. The equations for capacitance, ripple voltage, output voltage, and duty ratio are greatly simplified when expressed in terms of λ. Complicated quadratic equations for the output voltages of DC-DC converters operating in the discontinuous current mode are obsolete with the λ-ratio. The output voltage of the buck converter with discontinuous inductor current expressed as

$$V_o = \frac{RV_sD^2}{4Lf}\left[\sqrt{1+\frac{8Lf}{RD^2}}-1\right],$$

now becomes

$$V_o = \frac{DV_s}{\sqrt{\lambda}}.$$

The λ-ratio effectively normalizes the DC-DC converter equations with respect to the circuit inductance. Specific values of inductance are no longer required to fully observe and understand the behavior of a two-level DC-DC converter. In addition to the λ-ratio, a unifying theory of two-level DC-DC converters is presented. Instead of a separate analysis for each converter, a new theory is developed for a general converter with a two-level inductor-voltage waveform. The two-level voltage waveform leads to a general inductor-current waveform from which general equations for duty ratio, minimum inductance, capacitance, ripple voltage, and output voltage are derived. The circuit equations for the buck, boost, and buck/boost converters are merely subsets of the general equations. When the converter-specific voltages are substituted into the general equations, the formulas are immediately transformed to those of the specific converter.

It is hoped that the student will find the text easy to read, that the practicing engineer will find the information and programs readily applicable, and that the instructor of power electronics will welcome the pedagogical value that begins with analysis and ends with simulation. This book, therefore, is intended for the practicing engineer, the engineering student, and the professor who guided the former and now works with the latter.

Randall A. Shaffer, Ph.D.
Embry-Riddle Aeronautical University
Prescott, Arizona

Part

I

Electrical Energy and Power

The principles of electrical energy and power are presented in the first chapter to lay a foundation for the study of the fundamentals of power electronics. *Power Electronics* is the subdiscipline of electrical engineering that involves electrical energy conversion by the control of semiconductor devices. The subject encompasses simple rectifier circuits to sophisticated motor control systems.

The energy and power principles lead to mathematical descriptions of the crucially important circuit qualities of power delivery and efficiency.

1 Principles of Electrical Power

In This Chapter

- Introduction
- Electrical Power and Energy
- Power Computations
- Chapter Summary
- The MATLAB Toolbox
- Problems

INTRODUCTION

The subject of power electronics is the merger of the field of electrical power engineering and the technology of solid-state electronic devices. It is the discipline that involves the study, analysis, and design of circuits that convert electrical energy from one form to another.

Power electronic circuitry is needed because electrical devices require different forms of electricity. Some require an alternating current (AC) source, such as a transformer or AC motor; some require a direct current (DC) source, for example, the logic circuitry in a computer. When the appropriate form of electricity required by a device is not readily available, a power electronic circuit is needed to convert the available source into the form required by the device.

The study of power electronics begins with a review of the fundamental circuit laws and their application in the computation of electrical power.

ELECTRICAL POWER AND ENERGY

The instantaneous voltage *across* a linear resistance is governed by Ohm's law:

$$v(t) = i(t)R. \tag{1.1}$$

The instantaneous current *through* the resistance is

$$i(t) = \frac{v(t)}{R}. \tag{1.2}$$

The product of instantaneous voltage and instantaneous current is the instantaneous power absorbed by the resistance:

$$p(t) = v(t)i(t). \tag{1.3}$$

Equation 1.3 applies not only to resistances, but to any electrical device. The instantaneous power represents the rate at which energy is absorbed or the rate at which work is performed. The unit of power is the joule per second, or *watt* (W):

$$p(t) = \frac{dw(t)}{dt}. \tag{1.4}$$

The total energy absorbed by a device in a time interval from t_1 to t_2 is the integral of the instantaneous power:

$$W = \int_{t_1}^{t_2} p(t)\,dt. \tag{1.5}$$

Example 1.1 illustrates how the total absorbed energy is computed for a common incandescent light bulb over a specified time period.

EXAMPLE 1.1

A typical 100 W light bulb has a resistance of approximately 144 Ω. Find the rate at which energy is absorbed by the bulb when the voltage across it is $v(t) = 120\sqrt{2}\cos(120\pi t)$. How much energy is absorbed by the bulb in 1 hour?

Solution

The current through the bulb is

$$i(t) = \frac{v(t)}{R} = \frac{120\sqrt{2}\cos(120\pi t)}{144} = \frac{5\sqrt{2}}{6}\cos(120\pi t) \text{ A}$$

The instantaneous power absorbed by the bulb is

$$p(t) = v(t)i(t) = \left[120\sqrt{2}\cos(120\pi t) \right]\left[\frac{5\sqrt{2}\cos(120\pi t)}{6} \right] = 200\cos^2(120\pi t) \text{ W}.$$

The instantaneous power is plotted in Figure 1.1.

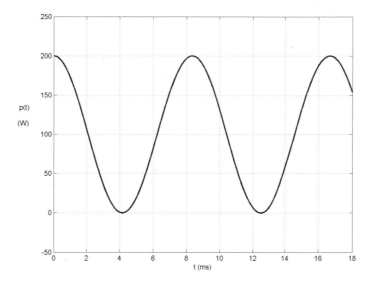

FIGURE 1.1   Instantaneous power absorbed by a 100 W light bulb.

There are 3600 seconds in an hour; therefore, the energy absorbed is

$$W = \int_0^{3600} 200\cos^2(120\pi t)\,dt = 200\left[\frac{t}{2} + \frac{1}{480}\sin(240\pi t) \right]_0^{3600}$$

$$W = (200)(1800) = 360,000 \text{ J}$$

Conclusion

When the voltage across a resistance is time varying, the rate at which energy is absorbed is also time varying. The instantaneous power in this example represents the rate at which energy is absorbed by a 100 W light bulb supplied by a typical 120 VAC (voltage alternating current) household source. As shown in Figure 1.1, the instantaneous power absorbed by a resistance is always positive, that is, a resistance cannot *store* energy but can only *consume* energy. The energy absorbed by the bulb is radiated as light and dissipated as heat. Unfortunately, about 95% of the energy is wasted as heat.

EXERCISE 1.1

Compute the cost to operate a 100 W light bulb for 8 hours a day for 30 days if the price of electricity is $0.12 for 3.6 million joules.

Answer

The cost is $2.88.

Average Values and Harmonics

As illustrated in Example 1.1, a periodic voltage applied across a linear resistance produces a periodic current of the same frequency. The same is true for all linear circuit elements. Power electronic circuits, however, contain nonlinear circuit elements that produce current components of frequencies in addition to that of the applied voltage source. Knowledge of these higher frequency components is essential in the design and analysis of power electronic circuits.

A periodic waveform has the property $x(t) = x(t + T)$. The fundamental period T is the shortest time interval over which the function $x(t)$ repeats itself. The fundamental frequency of the waveform is

$$f = \frac{1}{T}. \tag{1.6}$$

The unit for T is seconds; the unit for f is cycles per second, or hertz (Hz). A periodic waveform is expressed by a trigonometric Fourier series in the form of

$$x(t) = X_0 + \sum_{n=1}^{\infty} \left[a_n \cos(n\omega_0 t) + b_n \cos(n\omega_0 t) \right], \tag{1.7}$$

in which

$$a_n = \frac{2}{T} \int_0^T x(t)\cos(n\omega_0 t)\,dt \qquad (1.8)$$

and

$$b_n = \frac{2}{T} \int_0^T x(t)\sin(n\omega_0 t)\,dt. \qquad (1.9)$$

With the trigonometric identity,

$$a\cos\theta + b\sin\theta = \sqrt{a^2 + b^2}\ \sin\left[\theta + \arctan\left(\frac{a}{b}\right)\right], \qquad (1.10)$$

the Fourier series is more compactly expressed as

$$x(t) = X_0 + \sum_{n=1}^{\infty} x_n \sin(n\omega_0 t + \phi_n), \qquad (1.11)$$

in which

$$x_n = \sqrt{a_n^2 + b_n^2} \qquad (1.12)$$

and

$$\phi_n = \arctan\left(\frac{a_n}{b_n}\right). \qquad (1.13)$$

The terms inside the summation of Equation 1.11 are called the *harmonics* of $x(t)$. The first harmonic ($n = 1$) is the *fundamental component* of the waveform. The term ω_0 is the fundamental angular frequency with units of radians per second:

$$\omega_0 = \frac{2\pi}{T}. \qquad (1.14)$$

The term outside the summation in Equation 1.11 is the *average value* of $x(t)$ and is commonly known as the *DC component* of the waveform. The average value, or DC component, is computed from the integral equation

$$X_0 = \frac{1}{T} \int_0^T x(t)\, dt. \tag{1.15}$$

Since the integral represents the area between $x(t)$ and the time axis over one period, the average value is also determined from the bounded area divided by the period:

$$X_0 = \frac{\text{Area}}{T}. \tag{1.16}$$

In electrical applications, the function $x(t)$ in Equation 1.15 represents instantaneous voltage or current. The zero subscript indicates that the average value is the *zero-frequency* component of $x(t)$. For nonlinear circuits supplied by a sinusoidal source, the harmonics of $x(t)$ are the source frequency and the higher frequency components that are multiples of the source frequency.

The DC component is an important waveform parameter because of the ways that circuit elements respond to it. An inductor behaves as a short circuit to the average value, while a capacitor behaves as an open circuit. When a battery is charged, only the average value of the charging current causes the battery to absorb energy. In a DC power supply, it is the average value of the output voltage that is of interest. In the case of a DC motor, the average applied voltage produces an average current that generates torque.

EXAMPLE 1.2

Find the average value of $v(\theta) = v_n \cos(n\theta - \phi_n)$, in which n is a positive integer. The period of the waveform is $2\pi/n$.

Solution

Application of Equation 1.15 results in the integral equation:

$$V_0 = \frac{n}{2\pi} \int_0^{\frac{2\pi}{n}} v_n \cos(n\theta - \phi_n)\, d\theta.$$

Evaluation of the integral is as follows:

$$V_0 = \frac{n}{2\pi} \left[\frac{v_n}{n} \sin(n\theta - \phi_n) \right]_0^{\frac{2\pi}{n}} = \frac{v_n}{2\pi} \sin(n\theta - \phi_n) \Big]_0^{\frac{2\pi}{n}}.$$

$$V_0 = \left[\frac{v_n}{2\pi} \sin\left(2\pi - \phi_n\right) + \sin\left(\phi_n\right) \right].$$

$$V_0 = \frac{v_n}{2\pi}\left[-\sin\left(\phi_n\right) + \sin\left(\phi_n\right) \right] = 0.$$

Conclusion

Any sinusoid of any frequency and phase has an average value of zero over one period or any integral number of periods.

EXERCISE 1.2

Find the average value of $v(\theta) = V_m \sin\theta, \ 0 \leq \theta \leq \pi$.

Answer

$$V_0 = \frac{2V_m}{\pi}.$$

Power Absorbed in DC Circuits

In resistive DC circuits, Ohm's law becomes

$$V = IR, \tag{1.17}$$

and

$$I = \frac{V}{R}. \tag{1.18}$$

The power absorbed by the resistance is the product of voltage and current:

$$P = VI. \tag{1.19}$$

Substitution of Equation 1.18 into Equation 1.19 yields a DC power formula in terms of the voltage *across* the resistance:

$$P = \frac{V^2}{R}. \tag{1.20}$$

Similarly, substitution of Equation 1.17 into Equation 1.19 yields a DC power formula in terms of the current *through* the resistance:

$$P = I^2 R. \tag{1.21}$$

The next example illustrates an application of the DC power formulas.

EXAMPLE 1.3

A certain battery-operated device requires 3.0 V and draws a current of 40 mA. What series resistance is required to operate the device from a 12 V automobile battery? How much power is dissipated by the resistance?

Solution

The series resistance must drop the battery voltage by $12 - 3 = 9$ V for the device. The required resistance, therefore, is $9/0.04 = 225$ Ω. The dissipated power is $(0.04)^2(225) = 0.36$ W.

Conclusion

The difference between the source voltage and the load voltage determines the voltage across the resistance. The current through the series resistance is the same as that required by the load.

EXERCISE 1.3

The closest standard value of resistance to 225 Ω is 220 Ω. Standard resistance values have a tolerance of $\pm 5\%$ of the nominal value. If a standard 220 Ω resistor is used as the series resistance in Example 1.3, what is the possible range of load current?

Answer

39 mA $< i_{\text{LOAD}} <$ 42 mA.

Inductors and Capacitors in DC Circuits

The *impedance* of an inductor is

$$Z_L = j\omega L, \tag{1.22}$$

and the *inductive reactance* is

$$X_L = \omega L. \tag{1.23}$$

The *impedance* of a capacitor is

$$Z_C = \frac{1}{j\omega C},$$ (1.24)

and the *capacitive reactance* is

$$X_C = -\frac{1}{\omega C}.$$ (1.25)

Equations 1.22 through 1.25 reveal how the inductor and capacitor respond to a steady-state zero-frequency current: when $\omega = 0$, the inductive reactance is zero and the capacitive reactance is infinite. The impedance of the inductor thus presents itself as a short circuit to a DC component, and the impedance of a capacitor appears as an open circuit. A short circuit is defined as zero voltage with finite current; an open circuit is zero current with finite voltage. With zero voltage across an inductor, no power is dissipated. Similarly, no power is dissipated in a capacitor with zero current. Hence, neither the inductor nor the capacitor dissipates energy in a DC circuit. The next example illustrates how reactive elements affect the analysis of a DC circuit.

EXAMPLE 1.4

Find the steady-state power absorbed in each element in the circuit shown in Figure 1.2.

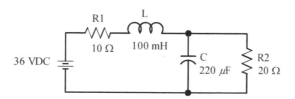

FIGURE 1.2   DC circuit with reactive elements.

Solution

To analyze a steady-state DC circuit, the inductor is replaced by a short circuit and the capacitor is replaced by an open circuit. The equivalent circuit is shown in Figure 1.3.

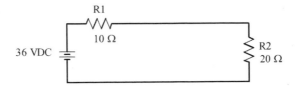

FIGURE 1.3   Equivalent steady-state DC circuit.

As shown in Figure 1.3, the circuit becomes a simple resistive circuit in steady-state. The two resistances are in series and therefore carry the same current, $36/30 = 1.2$ A. The power absorbed in $R1$ is $(1.2)^2(10) = 14.4$ W, and the power absorbed in $R2$ is $(1.2)^2(20) = 28.8$ W. The power delivered by the source is $(36)(1.2) = 43.2$ W. The power delivered by the source is also the sum of the powers absorbed by the resistors: $14.4 + 28.8 = 43.2$ W. The inductor and capacitor absorb no power in the steady-state condition.

Conclusion

In the steady-state circuit, the values of the inductances and capacitances are irrelevant in the computation of absorbed power because the energy delivered by the source is only dissipated in the resistive elements. The analysis of a steady-state DC circuit is simplified when all inductors are replaced with short circuits and all capacitors are replaced with open circuits.

EXERCISE 1.4

Find the power absorbed in each element in the circuit shown in Figure 1.4.

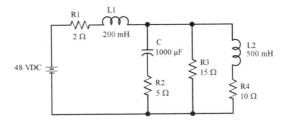

FIGURE 1.4   Circuit of Exercise 1.4.

Answers

$P_{R1} = 72$ W
$P_{R3} = 86.4$ W
$P_{R4} = 129.6$ W
$P_{48V} = 288$ W (supplied)
$P_{R2} = P_{L1} = P_{L2} = P_C = 0$.

Average Power

When the voltage and current waveforms of a device vary with time, the absorbed power is best described by its average value. The average value of power complies with the conservation of energy principle: the sum of all the average powers supplied equals the sum of all the average powers absorbed. The product of average power and a time interval represents the net work done, or net energy expended, during that interval. An electric bill is based on the average power used by the customer. Electric companies charge for each kilowatt-hour (KWH) of energy usage: the amount of energy expended when an average power of 1000 W is absorbed for 1 hour.

EXAMPLE 1.5

Determine the number of joules in 1 KWH.

Solution

1 KWH = (1000 J/s)(60 s/min)(60 min) = 3,600,000 J.

Conclusion

A very large number of joules are used in 1 hour at the rate of 1000 W. Even at the rate of 100 W, as in Example 1.1, the amount of joules used in 1 hour is 360,000. Because of the large number of joules, the KWH is a much more convenient unit of energy measurement for the use of electric power.

EXERCISE 1.5

Find the number of KWHs used by a 2-kilowatt air conditioning unit if operated 8 hours a day for 30 days. Calculate the cost of energy based on 12 cents per KWH.

Answer

480 KWH; the cost is $57.60.

Average Power in AC Circuits

The average power absorbed by a device is found upon application of the definition of the average value (Equation 1.15) to the instantaneous power (Equation 1.3):

$$P = \frac{1}{T} \int_0^T v(t)i(t)dt. \tag{1.26}$$

When the voltage and current in Equation 1.26 are both sinusoids of the same frequency, as in linear AC circuits, the average power is often called *real*, *active*, or *true* power. These adjectives are used because the average power represents the actual net energy rate of expenditure. The term *power* alone refers to average or real power unless otherwise specified.

EXAMPLE 1.6

Find the average power absorbed in an ideal inductor when the current through the inductance is periodic.

Solution

The voltage across an ideal inductor is

$$v(t) = L\frac{di(t)}{dt}. \tag{1.27}$$

Substitution of Equation 1.27 into Equation 1.26 produces the average power expression:

$$P = \frac{L}{T} \int_0^T \frac{di}{dt} i(t)dt.$$

Evaluation of the integral results in

$$\int_0^T i(t)\,di = \frac{1}{2}i^2(t)\Big|_0^T.$$

The average power is thus

$$P = \frac{L}{2T}\Big[i^2(T) - i^2(0) \Big].$$

Since the current is periodic, $i(0) = i(T)$. The average power absorbed, therefore, is zero.

Conclusion

Ideal inductors do not consume power; the net energy absorbed is zero.

EXERCISE 1.6

Prove that the average power absorbed in an ideal capacitor is zero when the current through the capacitor is

$$i(t) = C\frac{dv(t)}{dt},\qquad (1.28)$$

and the voltage across the capacitor is periodic.

Using MATLAB to Compute Average Values

The *trapz* function in MATLAB employs the trapezoidal method to approximate an integral. The function is readily used in the computation of average voltage, average current, and average power. The syntax of the function is

```
area = trapz(t, x)
```

In this instance, *trapz* computes the integral of **x** with respect to **t**. Both **x** and **t** must be vectors of the same length. The variable name *area* is arbitrary; it is a reminder that the integral represents the area bounded by **x(t)** and the horizontal axis. To compute the average value of a waveform, vector **t** is created over one period of the waveform. Upon application of the *trapz* function, the area is divided by the period to obtain the average value. The next example illustrates how to use *trapz* to compute an average voltage.

EXAMPLE 1.7

Use MATLAB to compute the average value of a full-wave rectified sine waveform that has a peak value of 170 V.

Solution

The solution is provided by the code in Listing 1.1. The program is available on the CD-ROM as file Example1_7.m in folder Chapter 1\Examples and Listings.

LISTING 1.1   Solution to Example 1.7

```
close all, clear all, clc
T = pi;
Vm = 170;
theta = linspace(0, T, 1024);
v = Vm*sin(theta);
VO = 1/T*trapz(theta, v)
```

Conclusion

The *close all* command closes all figures and any files that have been opened by MATLAB. The *clear all* command removes all variables and their values from the MATLAB workspace. The *clc* command removes all text from the command window. The first line in the solution code, in effect, wipes the slate clean.

The period of the rectified sine waveform is π. The variable name *pi* is permanently assigned the value of π in MATLAB. The function *linspace* creates a vector **theta** of 1024 equally spaced elements from zero to *pi*. Since **theta** is a vector, **sin(theta)** and **v** are also vectors of the same length as **theta**.

EXERCISE 1.7

Create an m-file in MATLAB with the code from Example 1.7 and verify that the average value computed is approximately 108 V. Verify that this result agrees with the result of Exercise 1.2 with $V_m = 170$ V.

Root-Mean-Square Values

The DC power formulas (Equations 1.20 and 1.21), along with the definition of average power (Equation 1.26), are used to define the root-mean-square (RMS) values of time-varying voltages and currents. The RMS values are used in the DC power formulas to calculate average power. By use of Ohm's law again, the average power absorbed by a resistance is expressed as

$$P = \frac{1}{T} \int_0^T \frac{v^2(t)}{R} dt. \tag{1.29}$$

With Equation 1.20 equated to Equation 1.29, the time-varying voltage is related to the DC voltage across the resistor:

$$\frac{1}{T} \int_0^T \frac{v^2(t)}{R} dt = \frac{V^2}{R}. \tag{1.30}$$

The solution of Equation 1.30 for the DC voltage is the *effective* value of the time-varying voltage:

$$V_{\text{RMS}} = \sqrt{\frac{1}{T} \int_0^T v^2(t)dt}.$$ (1.31)

The term *effective* is used because the RMS value has the same effect on average power as does a DC voltage of the same value. The effective value is more commonly referred to as the RMS value, a term from the mathematics of statistics. Similarly, the formula used to determine the effective value of current is

$$I_{\text{RMS}} = \sqrt{\frac{1}{T} \int_0^T i^2(t)dt}.$$ (1.32)

A DC voltage or current dissipates the same amount of power in a resistance as a time-varying voltage or current that has an RMS value equal to the DC value. For example, a DC voltage of 10 V across a 10 Ω resistance causes 10 W of power dissipation; a sinusoidal voltage across the resistance with an RMS value of 10 V also dissipates 10 W.

With RMS values, the DC power formulas are used to calculate the average power dissipated in a resistance when the voltage and current waveforms are periodic:

$$P = I_{\text{RMS}}^2 R.$$ (1.33)

$$P = \frac{V_{\text{RMS}}^2}{R}.$$ (1.34)

In practice, the term *RMS* is not often used for AC voltages; rather the term *VAC* is used, especially for ratings of transformers and AC motors. VAC literally means "voltage alternating current" but is better described as "alternating voltage *with* alternating current." The term *alternating* refers to the alternating polarity of a sinusoidal voltage and the alternating direction of a sinusoidal current.

EXAMPLE 1.8

Find the RMS value of the waveform from Example 1.2.

Solution

Application of the definition of RMS voltage yields

$$V_{\text{RMS}} = \sqrt{\frac{n}{2\pi} \int\limits_{0}^{\frac{2\pi}{n}} v_n^2 \cos^2\left(n\theta - \phi_n\right) d\theta}.$$

Evaluation of the integral and the RMS value proceeds as follows:

$$V_{\text{RMS}} = \sqrt{\frac{nv_n^2}{2\pi} \left[\frac{\theta}{2} + \frac{1}{4n} \sin\left(2n\theta - 2\phi_n\right)\right]_0^{\frac{2\pi}{n}}}.$$

$$V_{\text{RMS}} = \sqrt{\frac{nv_n^2}{2\pi} \left[\frac{\pi}{n} + \frac{1}{4n} \sin\left(4\pi - 2\phi_n\right) - \frac{1}{4n} \sin\left(-2\phi_n\right)\right]}.$$

$$V_{\text{RMS}} = \sqrt{\frac{nv_n^2}{2\pi} \left(\frac{\pi}{n}\right)}.$$

$$V_{\text{RMS}} = \frac{V_n}{\sqrt{2}}. \tag{1.35}$$

Conclusion

Any sinusoidal voltage of any frequency and phase has an RMS value that is equal to the peak value of the voltage divided by the square root of two. Similarly, the RMS value of a sinusoidal current is the peak current over the square root of two.

EXERCISE 1.8

The RMS value of a full-wave rectified sine waveform is the same as that of the non-rectified sine waveform. The RMS values are the same because the areas under the squared functions are the same. The average values of the two waveforms are different because the areas under the individual functions themselves are *not* the same. Find the RMS value of the waveform in Exercise 1.2 and prove that the RMS values of rectified and nonrectified sinusoids are the same.

Answer

$$V_{\text{RMS}} = \frac{V_m}{\sqrt{2}}.$$

The next example illustrates how a particular DC voltage and sinusoidal voltage dissipate the same average power in a resistance.

EXAMPLE 1.9

A 12 V battery placed across a 10 Ω resistance dissipates 14.4 W. What equivalent sinusoid dissipates the same average power?

Solution

To dissipate the same power, the RMS value of the sinusoid must also be 12 V. Therefore,

$$\frac{V_m}{\sqrt{2}} = 12,$$

and

$$V_m = 12\sqrt{2}.$$

Conclusion

A sinusoidal voltage with a peak value of $12\sqrt{2}$ V across a 10 Ω resistance dissipates the same power as when the resistance is placed across a 12 V battery.

EXERCISE 1.9

Calculate the power dissipated in a 10 Ω resistor if the sinusoidal current through the resistance has a peak value of 10 A.

Answer

$P = 500$ W.

Using MATLAB to Compute RMS Values

The *trapz* function is also used to compute RMS values as well as average values. For the RMS value, the integral of $x^2(t)$ is required rather than the integral of $x(t)$. The next example illustrates how to use *trapz* to compute RMS values.

EXAMPLE 1.10

Use MATLAB to verify the RMS value derivation of the full-wave rectified sine wave in Exercise 1.8 with $V_m = 170$ V.

Solution

ON THE CD

The solution is provided by the code in Listing 1.2. The code is available on the CD-ROM in folder Chapter 1\Examples and Listings, filename Example1_10.m.

LISTING 1.2   Solution to Example 1.10

```
close all, clear all, clc
T = pi;
Vm = 170;
theta = linspace(0, T, 1024);
v = Vm*sin(theta);
Vrms = sqrt(1/T*trapz(theta, v.^2))
```

Conclusion

The computation of the RMS value requires use of the .^ operator. This "dot-operator" squares each element in vector **v**.

EXERCISE 1.10

Find the RMS value of $v(t+T)=100t/T$, in which $T=2\pi$. Verify that the RMS value is the peak value divided by $\sqrt{3}$.

POWER COMPUTATIONS

A linear circuit is described as *sinusoidal* if all the voltages and currents in the circuit elements are sinusoids of one and the same frequency: that of the source or sources. For example, if a linear circuit is supplied by a 60 Hz voltage source, then all the voltages and currents in the circuit will be 60 Hz sinusoids. If a linear circuit is supplied by a nonsinusoidal source, then the circuit is described as *nonsinusoidal*. If a circuit is supplied by a single-frequency source and there arise voltages or currents of a frequency other than the source, then the circuit is described as *nonlinear* as well as nonsinusoidal.

The average power defined by Equation 1.26 applies to all circuits: linear, nonlinear, sinusoidal, and nonsinusoidal. However, more convenient formulae for the computation of average power are derived from Equation 1.26 in accordance with the type of circuit.

Sinusoidal Circuits

If the sinusoidal voltage across a circuit element is

$$v(t) = V_{\mathrm{m}} \sin(\omega t), \tag{1.36}$$

then the sinus͡ ͡ ͡   ugh the element is expressed as

$$I_\mathrm{m} \sin(\omega t \pm \phi). \tag{1.37}$$

ans/second) is related to the cyclical frequency

$$2\pi f, \tag{1.38}$$

ression

$$\tag{1.39}$$

ent has the same frequency as the volt-
with the voltage unless the element is
rent leads $(+\phi)$ or lags $(-\phi)$ the voltage
n of $\theta = \omega t$, Equations 1.36 and 1.37

$$\tag{1.40}$$

$$\tag{1.41}$$

Substitution of Equations 1.40
wer expression

$$\tag{1.42}$$

$$\tag{1.43}$$

the avei   d   ~y Equation 1.42 becomes

$$P = \frac{V_\mathrm{m} I_\mathrm{m}}{4\pi}\left[\int_0^{2\pi} \cos(\pm\phi)\,d\theta - \int_0^{2\pi} \cos(2\theta \pm \phi)\,d\theta\right]. \tag{1.44}$$

Since the first integral in Equation 1.44 is not a function of θ, the solution is

$$\int_0^{2\pi} \cos(\pm\phi)\,d\theta = 2\pi\cos(\phi). \tag{1.45}$$

Evaluation of the second integral in Equation 1.44 proceeds as follows:

$$\int_0^{2\pi} \cos(2\theta \pm \phi)\,d\theta = \frac{1}{2}\sin(2\theta \pm \phi)\Big]_0^{2\pi}. \tag{1.46}$$

$$\frac{1}{2}\Big[\sin(4\pi \pm \phi) - \sin(\pm\phi)\Big] = 0. \tag{1.47}$$

With Equations 1.45 through 1.47 substituted into Equation 1.44, the average power formula for sinusoidal circuits emerges as

$$P = \frac{V_m I_m}{2}\cos\phi. \tag{1.48}$$

An alternate form of Equation 1.48 is available in terms of the RMS values:

$$P = V_{RMS} I_{RMS} \cos\phi. \tag{1.49}$$

Equations 1.48 and 1.49 reveal that the average power absorbed in an element is dependent upon the angle by which the current leads or lags the voltage. If the current leads or lags by 90°, as in capacitors and inductors, the average power absorbed is zero. If the phase angle is zero, as in resistors, the average power is simply the product of RMS voltage and RMS current. A separate descriptive term for power is used in the special case of zero phase angle. The product of RMS voltage and RMS current is defined as the *apparent power*:

$$S = V_{RMS} I_{RMS}. \tag{1.50}$$

If the RMS voltage and current are measured for a two-terminal device, the product of the two RMS values "appears" to be the average power, hence the name "apparent" power. However, as Equation 1.49 dictates, the average power is only equal to the product of RMS voltage and current when $\phi = 0$, that is, only when the device is purely resistive. The units of apparent and average power are the same, namely, watts. However, the unit VA (volt-ampere) is used for apparent power to distinguish it from average power.

Apparent power is used as a specification for transformers and AC generators. The apparent power rating determines the maximum RMS source current. For example, if a 220 VAC source is specified by an apparent power of 5 kilovoltampere (kVA), then the maximum possible source current is $5000/220 \approx 22.7$ A. Furthermore, this maximum current draw is possible only when the load is purely resistive.

The apparent power is also useful as a comparison to the average power. The average power of a resistive-reactive load is readily compared to the case of a purely resistive load by the ratio of average to apparent power; this ratio is called the *power factor*:

$$pf = \frac{P}{S}. \tag{1.51}$$

As with average power, the apparent power and power factor definitions are universal; they apply to all circuits and periodic waveforms within the context of network theory. Formulae are derived from the definitions in accordance with the type of circuit. In the case of sinusoidal circuits, a convenient formula for power factor results from the ratio of Equation 1.49 to Equation 1.50:

$$pf = \cos\phi. \tag{1.52}$$

Equation 1.52 reveals that the power factor of a sinusoidal circuit is simply the cosine of the phase angle between the voltage and current, or the angle of the load impedance. The power factor represents the ratio of the average power of a reactive load to the average power of a resistive load. The next example illustrates power and power factor calculations for an inductive impedance.

EXAMPLE 1.11

Find the apparent power, power factor, and average power of a load impedance $Z_L = 10 + j7\ \Omega$ supplied by a 120 VAC source.

Solution

ON THE CD

The solution is provided by the code in Listing 1.3, which is available on the CD-ROM as file Example1_11.m in folder Chapter 1\Examples and Listings.

LISTING 1.3   Solution to Example 1.11

```
close all, clear all, clc
Vs = 120;
ZL = 10 + j*7;
R = real(ZL);
```

```
X = imag(ZL);
magZ = abs(ZL);
Is = Vs/magZ;
S = Vs*Is
phi = atan(X/R);
pf = cos(phi)
P = Vs*Is*pf
```

Conclusion

The letter j is permanently assigned the value of $\sqrt{-1}$ in MATLAB (as is the letter i). The resistive and reactive parts of the impedance are extracted by the *real* and *imag* functions, respectively. The magnitude of impedance is obtained with the absolute value function, *abs*. If the argument of *abs* is real, the absolute value is returned; if complex, the magnitude is returned. The computed results are $S = 1.2$ kVA, $pf = 0.82$, and $P = 966$ W.

Exercise 1.11

Find the apparent power, power factor, and average power of a load impedance $Z_L = 12 - j4 \ \Omega$ supplied by a 120 VAC source.

Answers

$S = 1.1$ kVA, $pf = 0.95$, and $P = 1080$ W.

Power Factor and Line Losses

Electrical loads with less-than-unity power factors draw larger currents than purely resistive loads for the same average power absorbed. For example, a 1.2 kW resistive load supplied by a 120 VAC source draws 10 A. If a reactive load with a power factor of 0.8 must absorb 1.2 kW, then the current drawn from the source is $10/0.8 = 12.5$ A. If supplied by distribution lines, the larger current required of the reactive load incurs larger losses in the supply lines than the resistive load. Power factor is thus an indicator of the energy lost in the distribution of electrical power. The losses tie directly to operational costs; hence load power factors are of vital interest to electric utility companies.

Electrical loads are typically inductive, largely because of transformers and AC motors. Inductive loads, fortunately, have power factors that can be improved by the addition of a parallel, or *shunt*, capacitor across the load. With a particular value of capacitance, the power factor of the combined load-capacitor impedance is unity. An inductive load with shunt capacitance is illustrated in Figure 1.5.

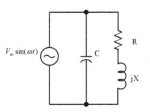

FIGURE 1.5   Inductive load with shunt capacitance.

Unity load power factor requires an impedance (or admittance) that is purely resistive; the imaginary part must be zero. If the impedance of the load is $Z_1 = R + jX$, then the admittance of the combined load with shunt capacitance is

$$Y_L = \frac{1}{R+jX} + j\omega C. \tag{1.53}$$

Multiplication of the rational part of the admittance by its complex conjugate results in

$$Y_L = \frac{R-jX}{R^2 + X^2} + j\omega C. \tag{1.54}$$

Combination of the real and imaginary parts of Equation 1.54 produces the resistive and reactive parts of the admittance:

$$Y_L = \frac{R}{R^2 + X^2} + j\left[\omega C - \frac{X}{R^2 + X^2}\right]. \tag{1.55}$$

For unity power factor, the imaginary part of Equation 1.55 must be zero:

$$\omega C - \frac{X}{R^2 + X^2} = 0. \tag{1.56}$$

The solution of Equation 1.56 for C is the value of the *power factor correction* (PFC) capacitor:

$$C = \frac{X}{\omega\left(R^2 + X^2\right)}. \tag{1.57}$$

With the value of capacitance expressed by Equation 1.57, the combined load impedance becomes

$$Z = \frac{R^2 + X^2}{R}.$$

(1.58)

The next example illustrates an application of Equations 1.57 and 1.58.

EXAMPLE 1.12

Find the value of shunt capacitance to correct the power factor of the load in Example 1.11 to unity if the source frequency is 60 Hz. Find the current absorbed by the combined load and the kVA rating of the source with the capacitor in the circuit.

Solution

The value of capacitance is readily determined from Equation 1.57:

$$C = \frac{7}{120\pi(10^2 + 7^2)} \approx 125 \ \mu\text{F}.$$

With this value of capacitance across the load, the combined impedance is

$$Z_c = \frac{10^2 + 7^2}{10} = 14.9 \ \Omega.$$

The current drawn from the source is thus $120/14.9 = 8.05$ A, and the apparent power of the source is $(120)(8.05) = 966$ VA.

Conclusion

A 125 μF capacitor placed in parallel with a $10 + j7\Omega$ load impedance results in a power factor of unity for the combined impedance. The capacitor causes an *increase* in the total load impedance. As a result of the increased impedance, the current drawn from the source is reduced. Thus, loads with higher power factors draw less current and require less-costly sources with lower apparent power ratings. With unity power factor, the apparent power of the source and the average power absorbed by the load are equal, consistent with the principle of zero absorbed average power in ideal capacitors.

EXERCISE 1.12

Find the apparent power of a 230 VAC source that supplies a load impedance $Z_L = 20 + j15\Omega$. Find the apparent power of the source after the load power factor is corrected to unity.

Answers

$S = 2.1$ kVA and $S_c = 1.7$ kVA.

Reactive Power

The PFC method discussed in the previous section has a notable disadvantage: the resistive and reactive parts of the load must be known in order to compute the value of capacitance. The values of these parameters are usually known only when PFC is an integral part of a system design. When a PFC capacitor is retrofitted, a PFC method that relies on circuit measurements is more practical.

To facilitate the PFC process with measurements, a mathematical counterpart to the average power is introduced as the *reactive power* and defined as

$$Q = V_{RMS}I_{RMS}\sin\phi. \tag{1.59}$$

Unlike average power, reactive power does not represent a physical process in a circuit; it exists solely as a mathematical entity. Reactive power does, however, make the PFC process quite simple. The unit of reactive power is the same as apparent power and average power: the volt-ampere or watt. However, in order to distinguish Q from S and P, the unit *VAr* is used: volt-ampere reactive (pronounced "var"). Equations 1.49, 1.50, and 1.59 suggest the geometric relationship of the real, reactive, and apparent powers as illustrated in Figure 1.6.

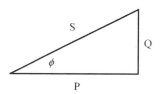

FIGURE 1.6   Geometric relationship of the real, reactive, and apparent powers.

In the case of a pure resistance, $\phi = 0$ in Equation 1.59 and thus $Q = 0$. Unity power factor therefore dictates that the total reactive power of a load be zero. In a pure inductance, $\phi = 90°$ and Q is positive. In a pure capacitance, $\phi = -90°$ and Q is negative. Herein is the utility of reactive power: a capacitor with negative Q can neutralize an inductive load with positive Q.

Apparent power and average power are readily measurable with laboratory instruments. When apparent power is measured, the voltmeter and ammeter sections of the instrument determine the respective RMS values, after which the product of RMS voltage and current is computed. Average power is measured with a *wattmeter*. With S and P measured, the reactive power is readily computed as

$$Q = \sqrt{S^2 - P^2}. \tag{1.60}$$

Reactive power is also computed with RMS voltage and current values in a manner similar to average power. With P replaced by Q and R replaced by X in Equations 1.33 and 1.34, the reactive power formulas include

$$Q = I_{\text{RMS}}^2 X, \tag{1.61}$$

and

$$Q = \frac{V_{\text{RMS}}^2}{X}, \tag{1.62}$$

in which X is the inductive or capacitive reactance. The RMS value in Equation 1.61 must be that of the current *through* the reactive element; the RMS value in Equation 1.62 must be that of the voltage *across* the reactive element.

EXAMPLE 1.13

Calculate the reactive power an inductive load $Z_{\text{L}} = 10 + \mathbf{j}6 \, \Omega$ supplied by a 120 VAC, 60 Hz voltage source.

Solution

The direct way to find Q for this example is to first compute the load current since this current is also the current through the reactive element $\mathbf{j}6$:

$$I_{\text{RMS}} = \frac{120}{\sqrt{10^2 + 6^2}} = 10.3 \text{ A.}$$

With Equation 1.61, the Q of the inductor is determined as

$$Q = I_{RMS}^2 X = (10.3)^2 (6) = 635 \text{ VAr.}$$

An alternate way to compute Q is to use Equation 1.59 and the angle of the impedance:

$$\phi = \arctan\left(\frac{X}{R}\right) = \arctan(0.6) = 0.54 \text{ radian.}$$

$$Q = V_{RMS} I_{RMS} \sin\phi = (120)(10.3)\sin(0.54) = 635 \text{ VAr.}$$

Still another way to compute Q is to find the voltage across the reactive element and then use Equation 1.62. The voltage across the inductance is found by voltage division:

$$V_L = 120\frac{6}{\sqrt{10^2 + 6^2}} = 61.7 \text{ VAC.}$$

Once again, the Q of the load is

$$Q = \frac{V_{RMS}^2}{X} = \frac{61.7^2}{6} = 635 \text{ VAr.}$$

Conclusion

As with average power, there are different ways to compute the reactive power of a load. The method used depends largely on whether the load voltage or load current is known.

EXERCISE 1.13

Find the value of shunt capacitance for the load of Example 1.13 that will neutralize the reactive power.

Answer

$C = 117 \, \mu\text{F.}$

Neutralizing Reactive Power

As stated earlier, the negative reactive power of a capacitance is used to neutralize the positive reactive power of an inductive load. The algebraic sum of the load reactive power and the reactive power of the capacitor must be zero:

$$Q_L + Q_C = 0. \qquad (1.63)$$

Since the capacitor and load are in parallel, the load voltage determines the capacitor reactive power by Equation 1.62. Substitution of Equation 1.62 into Equation 1.63 produces

$$Q_L + \frac{V_{RMS}^2}{X_C} = 0. \qquad (1.64)$$

Substitution of Equation 1.25 into Equation 1.64 yields

$$Q_L - \omega C V_{RMS}^2 = 0. \qquad (1.65)$$

The solution of Equation 1.65 for C is the capacitance required to neutralize the Q of the load and correct the power factor to unity:

$$C = \frac{Q_L}{\omega V_{RMS}^2}. \qquad (1.66)$$

Equation 1.66 yields the same value of capacitance as Equation 1.59. The advantage to Equation 1.66 is that the load parameters do not need to be determined or known explicitly; only the measurements of S, P, and V_{RMS} are required.

EXAMPLE 1.14

A load with a poor power factor has a measured voltage of 240 VAC and draws a measured RMS current of 27 A. The absorbed power measured by a wattmeter is 3888 W. If the source frequency is 60 Hz, find the value of shunt capacitance that corrects the power factor. Find the current drawn from the source with the capacitor in the circuit if the source impedance is negligible.

Solution

ON THE CD

The solution is provided by the code in Listing 1.4. The code is available as file Example 1_14.m in the folder Chapter 1\Examples and Listings on the CD-ROM.

LISTING 1.4 Solution Code for Example 1.14

```
close all, clear all, clc
f = 60;
w = 2*pi*f;
Vrms = 240;
Irms = 27;
```

```
P = 3888;
S = Vrms*Irms;
Q = sqrt(S^2 - P^2);
C = Q/(w*Vrms^2)
Irms = P/Vrms
```

Conclusion

The computed value of capacitance is 239 µF. With this capacitor in the circuit, the load draws only 16.2 A from the source. Since the capacitor consumes no average power and the voltage across the load is unchanged because of negligible source impedance, the power absorbed by the load does not change. The effect of the shunt capacitor is thus threefold: the total reactive power seen by the source is zero, the power factor seen by the source is unity, and the current drawn from the source is substantially decreased.

EXERCISE 1.14

Find the value of shunt capacitance to correct the power factor of a load with measurements as follows: $P = 10$ KW, $pf = 0.72$, $I_{RMS} = 42$ A, and $f = 60$ Hz. Compute the current drawn from the load with the capacitor in the circuit. Assume that the source impedance is negligible.

Answers

$C = 234$ µF and $I_{RMS} = 30.2$ A.

Practical Applications of Power Factor Correction

In practical applications, there is series impedance between the source and the load. This impedance represents the impedance of the source and the impedance of the distribution lines between the source and the load. The series load impedances establish a voltage divider that reduces the source voltage at the load. However, since the PFC capacitance increases the load impedance, the load voltage increases with the capacitor in the circuit. The increased load voltage subsequently increases the absorbed load power.

The next example illustrates the effects of PFC in a practical situation where the resistance and reactance of the distribution line are included in the computations.

EXAMPLE 1.15

Find the power absorbed by the load, the power lost in the line, and the efficiency of the circuit shown in Figure 1.7. Find the value of the PFC capacitor. Recalculate the load power, line losses, and efficiency with the shunt capacitor in the circuit.

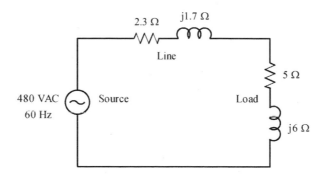

FIGURE 1.7   Power distribution circuit with line impedance.

Solution

The solution is provided by the program in Listing 1.5. The program is available as file Example1_15.m on the CD-ROM in folder Chapter 1\Examples and Listings.

LISTING 1.5   Solution Program for Example 1.15

```
close all, clear all, clc
f = 60;
w = 2*pi*f;
Vs = 480;
RT = 2.3;
XT = 1.7;
RL = 5;
XL = 6;
ZT = RT + j*XT;
ZL = RL + j*XL;
Zs = ZT + ZL;          % Total circuit impedance
mZs = abs(Zs);
mZL = abs(ZL);
Is = Vs/mZs;
PL = Is^2*RL           % Load power
PT = Is^2*RT           % Transmission line losses
eta = PL/(PT + PL)     % Efficiency
QL = Is^2*XL;
VL = Vs*mZL/mZs;
C = QL/(w*VL^2)        % PFC capacitance
Zr = (RL^2 + XL^2)/RL;
ZsC = Zr + ZT;
```

```
mZsC = abs(ZsC);
VLC = Vs*Zr/mZsC;
ILC = VLC/mZL;
PLC = ILC^2*RL                    % Load power with PFC
IsC = Vs/mZsC;
PTC = IsC^2*RT           % Line losses with PFC
etac = PLC/(PLC + PTC)% Efficiency with PFC
```

Conclusion

The load and line powers are 10.2 kW and 4.7 kW, respectively, and result in an efficiency of 68.5%. The addition of the shunt capacitor (261 μF) to the circuit has remarkable results. The power lost in the distribution line is greatly reduced to 2.49 kW, and the power delivered to the load is significantly increased to 13.2 kW. The efficiency, consequently, is a much improved 84.1%.

EXERCISE 1.15

Repeat Example 1.15 with a load impedance of $Z_L = 8 + j4\ \Omega$.

Answers

$P_L = 13.3$ KW
$P_T = 3.82$ KW
$\eta = 77.7\%$
$C = 133\ \mu F$
$P_{L,C} = 14.9$ KW
$P_{T,C} = 3.44$ KW
$\eta = 81.3\%$.

Nonlinear Circuits with Sinusoidal Sources

The term *nonlinear* refers to circuits that draw nonsinusoidal current from a sinusoidal voltage source. The source voltage is composed of one frequency only; the source current, however, is composed of possibly an infinite number of frequencies. The sinusoidal source voltage is expressed as

$$v(\theta) = V_m \sin \theta, \tag{1.67}$$

in which

$$\theta = \omega t. \tag{1.68}$$

If the source supplies a current $i(\theta)$, then in accordance with Equation 1.26, the average power delivered by the source is

$$P = \frac{1}{T}\int_0^T v(\theta)i(\theta)d\theta, \tag{1.69}$$

in which T has units of radians. Substitution of Equation 1.67 into Equation 1.69 produces

$$P = \frac{V_m}{T}\int_0^T i(\theta)\sin\theta\, d\theta. \tag{1.70}$$

A comparison of Equation 1.70 to Equation 1.9 reveals a relationship between the average power of a nonlinear load and the sine coefficients of the Fourier series expansion of $i(\theta)$. Substitution of $n=1$ and $\theta=\omega t$ into Equation 1.9 produces the *fundamental* sine coefficient expression

$$b_1 = \frac{2}{T}\int_0^T i(\theta)\sin\theta\, d\theta. \tag{1.71}$$

With the relationships of Equations 1.70 and 1.71, the formula for the average power of a nonlinear load in terms of the fundamental sine coefficient is

$$P = \frac{V_m b_1}{2}. \tag{1.72}$$

Although Equation 1.72 is a convenient formula with which to compute the average power of a nonlinear load, a more insightful expression is obtained when the current is expressed in the form of the compact Fourier series

$$i(\theta) = I_0 + \sum_{n=1}^{\infty} I_n \sin(n\theta - \phi_n). \tag{1.73}$$

It appears that, upon substitution of Equation 1.73 into Equation 1.70, the average power integral consists of an infinite number of terms. However, all but one of those terms is zero because of the orthogonal property of sinusoids of unlike frequencies. Two functions $x(\theta)$ and $y(\theta)$ are orthogonal over an interval θ_1 to θ_2 if their product satisfies the integral

$$\int_{\theta_1}^{\theta_2} x(\theta) y(\theta) d\theta = 0. \tag{1.74}$$

If $x(\theta)$ and $y(\theta)$ are sinusoids of different frequencies and the interval is one period of the lowest frequency component, then those sinusoids are orthogonal. Hence, the only current component that contributes to absorbed power is the fundamental: the case of $n = 1$ in Equation 1.73:

$$i_1(\theta) = I_1 \sin(\theta - \phi_1). \tag{1.75}$$

Substitution of Equation 1.75 into Equation 1.70 results in the average power expression

$$P = \frac{V_m I_1}{2\pi} \int_0^{2\pi} \sin\theta \sin(\theta - \phi_1) d\theta. \tag{1.76}$$

Equation 1.76 has the same form as Equation 1.42. The solution to the integral, therefore, has the same form as Equation 1.48:

$$P = \frac{V_m I_1}{2} \cos\phi_1. \tag{1.77}$$

Expressed in terms of the RMS values, Equation 1.77 becomes

$$P = V_{RMS} I_{1,RMS} \cos\phi_1, \tag{1.78}$$

which is similar in form to Equation 1.49. Equation 1.78 indicates that the average power absorbed by a circuit that draws nonsinusoidal current from a sinusoidal source is dependent upon the RMS source voltage and the parameters of the fundamental component of current, namely, the RMS value of the fundamental current and the phase angle between the source voltage and the fundamental current.

The power factor is the ratio of Equation 1.78 to the apparent power:

$$pf = \frac{I_{1,RMS}}{I_{RMS}} \cos\phi_1. \tag{1.79}$$

Equation 1.79 indicates that the power factor of a nonlinear load is composed of two parts: (a) the ratio of the RMS value of the fundamental component to the RMS value of current and (b) the cosine of the angle between the source voltage and the fundamental component. The first factor in Equation 1.79 is due to harmonics in the current; the second is due to a phase difference between the source voltage and the fundamental current. The ratio of RMS currents is defined as the *distortion factor* (DF):

$$DF = \frac{I_{1,\text{RMS}}}{I_{\text{RMS}}}.$$ (1.80)

Angle ϕ_1 in Equation 1.79 is the angle by which the fundamental current is *displaced* from the source voltage; hence, the cosine of this angle is called the *displacement power factor* (DPF) or, more succinctly, the displacement factor:

$$DPF = \cos\phi_1.$$ (1.81)

The power factor of a nonlinear load, therefore, is the product of the distortion factor and the displacement factor:

$$pf = (DF)(DPF).$$ (1.82)

The next example illustrates the average power, power factor, and related computations for a nonlinear circuit supplied by a sinusoidal source.

EXAMPLE 1.16

A nonlinear circuit is supplied by a voltage source $v(\theta) = 120\sqrt{2}\sin\theta$ V and draws the nonsinusoidal current

$$i(\theta) = i(\theta + 2\pi) = \begin{cases} 10, & 0 \le \theta \le \pi \\ 0, & \pi < \theta \le 2\pi \end{cases}.$$

Find the apparent and average powers and the distortion, displacement, and power factors.

Solution

The RMS voltage is

$$V_{\text{RMS}} = \frac{120\sqrt{2}}{\sqrt{2}} = 120 \text{ V},$$

and the RMS current is

$$I_{RMS} = \sqrt{\frac{1}{2\pi} \int_0^\pi (10)^2 \, d\theta} = \sqrt{\frac{50}{\pi} \theta \Big|_0^\pi} = \sqrt{50} = 7.07 \text{ A.}$$

The apparent power is

$$S = V_{RMS} I_{RMS} = (120)(7.07) = 849 \text{ VA.}$$

The Fourier coefficients of the fundamental current are

$$a_1 = \frac{2}{T} \int_0^T i(\theta)\cos(\theta) \, d\theta = \frac{1}{\pi} \int_0^\pi (10)\cos\theta \, d\theta = \frac{10}{\pi}\sin\theta \Big]_0^\pi = 0$$

and

$$b_1 = \frac{2}{T} \int_0^T i(\theta)\sin(\theta) \, d\theta = \frac{1}{\pi} \int_0^\pi (10)\sin\theta \, d\theta = -\frac{10}{\pi}\cos\theta \Big]_0^\pi = \frac{20}{\pi}.$$

The magnitude of the fundamental component of current is

$$I_1 = \sqrt{a_1^2 + b_1^2} = \frac{20}{\pi} \text{ A,}$$

and the RMS value is

$$I_{1,RMS} = \frac{20}{\pi\sqrt{2}} = 4.50 \text{ A.}$$

The phase angle of the fundamental current is

$$\phi_1 = \arctan\left(\frac{a_1}{b_1}\right) = \arctan\left(\frac{0}{b_1}\right) = 0.$$

Since the phase angle of the source voltage is also zero, the average power is

$$P = V_{RMS} \, I_{1,RMS} \cos\phi_1 = (120)(4.50)\cos(0°) = 540 \text{ W,}$$

and the power factor is

$$pf = \frac{P}{S} = \frac{540}{849} = 0.64.$$

The distortion factor is

$$DF = \frac{I_{1,\text{RMS}}}{I_{\text{RMS}}} = \frac{4.50}{7.07} = 0.64,$$

and the displacement factor is

$$DPF = \cos\phi_1 = \cos 0° = 1.$$

Conclusion

In this example, the fundamental current is in phase with the source voltage. The less-than-unity power factor is due entirely to current harmonics above the fundamental component.

EXERCISE 1.16

Find the apparent and average powers and the distortion, displacement, and power factors of a circuit with sinusoidal voltage $v(\theta) = 220\sqrt{2}\sin\theta$ V and nonsinusoidal current

$$i(\theta) = i(\theta + 2\pi) = \begin{cases} 30\sin\theta, & 0 \leq \theta \leq \pi \\ 0, & \text{otherwise} \end{cases}.$$

Answers

$S = 3.30$ kVA
$P = 2.33$ kW
$DF = 0.71$
$DPF = 1$
$pf = 0.71.$

Orthogonal Functions and RMS Values

Since the periodic current drawn by a nonlinear load is often expressed as a Fourier series, it is advantageous to use the properties of the series to facilitate power computations. The property that simplifies power computations is the orthogonal property of sinusoids of unlike frequencies.

The orthogonal property of sinusoids provides a very convenient way to compute the RMS value of a waveform that is the sum of sinusoids of unlike frequencies. For example, if a current is composed of two sinusoids as

$$i(\theta) = I_1 \cos n\theta + I_2 \cos m\theta, \quad n \neq m, \tag{1.83}$$

then the square of the RMS value of Equation 1.83 is computed as follows:

$$I_{\text{RMS}}^2 = \frac{1}{T}\int_0^T \left[I_1 \cos n\theta + I_2 \cos m\theta \right]^2 d\theta. \tag{1.84}$$

$$I_{\text{RMS}}^2 = \frac{1}{T}\int_0^T \left[I_1^2 \cos^2 n\theta + 2 I_1 I_2 \cos n\theta \cos m\theta + I_2^2 \cos^2 m\theta \right] d\theta. \tag{1.85}$$

$$I_{\text{RMS}}^2 = \frac{1}{T}\int_0^T I_1^2 \cos^2 n\theta d\theta + \frac{2 I_1 I_2}{T}\int_0^T \cos n\theta \cos m\theta d\theta + \frac{1}{T}\int_0^T I_2^2 \cos^2 m\theta d\theta. \tag{1.86}$$

The second integral in Equation 1.86 evaluates to zero because the sinusoids are orthogonal. The first and third integrals are simply the squares of the RMS values of the individual sinusoids:

$$\frac{1}{T}\int_0^T I_1^2 \cos^2 n\theta \, d\theta = I_{1,\text{RMS}}^2 = \frac{1}{2}I_1^2. \tag{1.87}$$

$$\frac{1}{T}\int_0^T I_2^2 \cos^2 m\theta \, d\theta = I_{2,\text{RMS}}^2 = \frac{1}{2}I_2^2. \tag{1.88}$$

With Equations 1.87 and 1.88, the RMS value of Equation 1.83 is simply

$$I_{\text{RMS}} = \sqrt{I_{1,\text{RMS}}^2 + I_{2,\text{RMS}}^2} = \sqrt{\frac{1}{2}I_1^2 + \frac{1}{2}I_2^2}. \tag{1.89}$$

Equation 1.89 reveals that the RMS value of a sum of sinusoids of different frequencies is the square root of the sum of the squares of the RMS values of the individual sinusoids. This result is extended to any number of sinusoids of unlike frequencies. Furthermore, the result is applicable to a function that is the sum of orthogonal functions of any type. If a particular function is expressed as

$$x(t) = x_1(t) + x_2(t) + \cdots + x_n(t), \tag{1.90}$$

in which $x_1(t)$, $x_2(t)$, …, $x_n(t)$ are all mutually orthogonal functions, then the RMS value of $x(t)$ is

$$X_{\text{RMS}} = \sqrt{X_{1,\text{RMS}}^2 + X_{2,\text{RMS}}^2 + \cdots + X_{n,\text{RMS}}^2}. \tag{1.91}$$

The next example illustrates the RMS value computation of a sum of orthogonal functions.

EXAMPLE 1.17

Find the RMS value of

$$i(\theta) = 10 + 20\cos(\theta - 30°) + 15\cos(2\theta - 45°) + 5\cos(3\theta - 60°).$$

Solution

Application of Equation 1.91 results in

$$I_{RMS} = \sqrt{(10)^2 + \frac{1}{2}(20)^2 + \frac{1}{2}(15)^2 + \frac{1}{2}(5)^2} = 20.6 \text{ A}.$$

Conclusion

The RMS value of a constant is simply the constant itself. The DC component of current is orthogonal to the harmonics, as it can be treated as a sinusoid of zero frequency: $I_0 = 10\cos(0)$.

EXERCISE 1.17

Find the RMS value of

$$i(\theta) = 1 + 4\cos(\theta - 15°) + 3\sin(\theta - 15°) + 2\cos(2\theta - 15°).$$

Answer

$I_{RMS} = 3.94$ A.

Using MATLAB to Compute RMS Values of Orthogonal Functions

The RMS value of a waveform expressed as Equation 1.90 is readily computed in MATLAB with the magnitudes of the harmonics placed in a vector. The code segment in Listing 1.6 illustrates the use of the *sum* function to compute the RMS value of the waveform in Example 1.17.

LISTING 1.6   Code to Compute the RMS Value of $i(\theta)$ in Example 1.17

```
close all, clear all, clc
Io = 10;
In = [20 15 5];
Irms = sqrt(Io^2 + 0.5*sum(In.^2))
```

The next example illustrates how to use MATLAB to compute the RMS value of a waveform with known Fourier coefficients.

EXAMPLE 1.18

Use MATLAB to compute the RMS value of the voltage waveform

$$v(\theta) = \frac{2V_m}{\pi} + \sum_{n=2,4,\ldots}^{8} \frac{4V_m}{\pi(n^2-1)}\cos(n\theta+\pi),$$

in which $V_m = 170$ V. Plot the waveform over two cycles of the fundamental frequency.

Solution

ON THE CD

The solution is provided by the code in Listing 1.7. The program is available on the CD-ROM as Example1_18.m in Chapter 1\Examples and Listings.

LISTING 1.7 Solution to Example 1.18

```
close all, clear all, clc
N = 8;
n = 2:2:N;
t = linspace(0, 4*pi, 1024);
Vm = 170;
V0 = 2*Vm/pi;
Vn =4*Vm./(pi*(n.^2 - 1));
theta - n'*t;
phi = pi*ones(size(theta));
v = V0 + Vn*cos(theta - phi);
plot(t, v), grid
Vrms = sqrt(V0^2 + 0.5*sum(Vn.^2))
```

Conclusion

The fundamental period of the waveform is 2π; hence, a linear space from 0 to 4π yields two cycles of the waveform. The dimensions of vector **t** are 1×1024; the dimensions of vector **n** are 1×4. The product **n'*t** produces the matrix **theta** with dimensions 4×1024. The apostrophe (') operator performs the transpose of a matrix or vector and is required to validate the matrix product. The **ones** vector is the same size as **theta** and forms the matrix **phi** with the same dimensions as **theta** to validate the subtraction operation. The computed RMS value of the waveform is 120 V. The waveform is plotted in Figure 1.8.

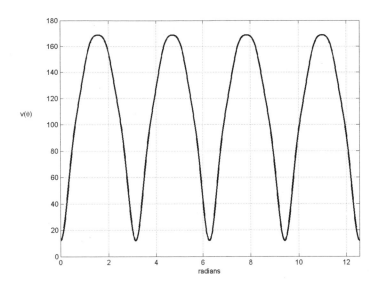

FIGURE 1.8   Waveform from Example 1.18.

Exercise 1.18

Use MATLAB to compute the RMS value of the voltage waveform

$$v(\theta) = \left(\frac{\pi - \alpha}{2\pi}\right)V_m + \sum_{n=1,3,\ldots}^{9} \frac{2V_m}{n\pi}\sin\left[n\left(\frac{\pi - \alpha}{2}\right)\right]\cos\left[n\left(\theta - \frac{\pi + \alpha}{2}\right)\right]$$

in which $V_m = 170$ V and $\alpha = \pi/6$. Plot the waveform over two cycles of the fundamental frequency.

Answer

$V_{RMS} = 104$ V.

Reactive Power in Nonlinear Circuits with Sinusoidal Sources

As demonstrated with sinusoidal circuits, the concept of reactive power is useful in the process of PFC with shunt capacitance. In the case of nonlinear circuits, a similar definition of reactive power is also available for power factor improvement with a shunt capacitor. As the reactive power expressed by Equation 1.59 defines the mathematical counterpart to the average power expressed by Equation 1.49, a mathematical counterpart to Equation 1.78 defines the reactive power in nonlinear circuits:

$$Q = -V_{\text{RMS}} \, I_{1,\text{RMS}} \sin \phi_1. \tag{1.92}$$

Equations 1.78 and 1.92 suggest the geometric relationship illustrated in Figure 1.9.

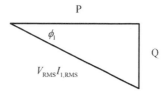

FIGURE 1.9   Geometric relationship suggested by Equations 1.78 and 1.92.

Reactive volt-amperes arise in a nonsinusoidal circuit when there is a phase difference between the source voltage and the fundamental component of the current drawn by the nonlinear load. If the fundamental component of current lags the source voltage, that is, if ϕ_1 is negative, then the reactive power of the circuit is positive. The positive reactive power can then be neutralized by the addition of shunt capacitance. If ϕ_1 is positive, however, the power factor cannot be improved by shunt capacitance; the addition of a shunt capacitor only further erodes the power factor.

In sinusoidal circuits, the apparent power is two-dimensional; the two components are the average and reactive powers. In nonlinear circuits supplied by a sinusoidal source, the apparent power is three-dimensional; the third component is called the *distortive* power:

$$S = \sqrt{P^2 + Q^2 + D^2}. \tag{1.93}$$

The geometry suggested by Equation 1.93 is illustrated in Figure 1.10.

FIGURE 1.10   Geometry suggested by Equation 1.93.

The expression for the distortive power is determined from the solution of Equation 1.93 for D:

$$D = \sqrt{S^2 - P^2 - Q^2}.\tag{1.94}$$

Upon substitution of the respective definitions for S, P, and Q into Equation 1.94, the derivation of the distortive power formula proceeds as follows:

$$D = \sqrt{V_{RMS}^2 I_{RMS}^2 - V_{RMS}^2 I_{1,RMS}^2 \cos^2 \phi_1 - V_{RMS}^2 I_{RMS}^2 \sin^2 \phi_1}$$

$$D = \sqrt{V_{RMS}^2 I_{RMS}^2 - V_{RMS}^2 I_{1,RMS}^2 (\cos^2 \phi_1 + \sin^2 \phi_1)}$$

$$D = \sqrt{V_{RMS}^2 I_{RMS}^2 - V_{RMS}^2 I_{1,RMS}^2}$$

$$D = V_{RMS} \sqrt{I_{RMS}^2 - I_{1,RMS}^2}.\tag{1.95}$$

The quantity $\sqrt{I_{RMS}^2 - I_{1,RMS}^2}$ in Equation 1.95 is the RMS value of all the current harmonics except the fundamental. The distortive power therefore represents that component of the apparent power that is due to the presence of current harmonics above the fundamental. The unit of distortive power is the *VAd*: volt-ampere distortive. The next example illustrates the computation of reactive and distortive power in a nonlinear circuit.

EXAMPLE 1.19

Find the reactive and distortive powers of the nonlinear circuit from Example 1.16. Verify the apparent power computation with Equation 1.93.

Solution

The reactive power is zero because the fundamental component of current is in phase with the source voltage, that is, $\phi_1 = 0$. The distortive power is

$$D = V_{RMS} \sqrt{I_{RMS}^2 - I_{1,RMS}^2} = 120 \sqrt{7.07^2 - 6.37^2} = 654 \text{ VAd}$$

and the apparent power is

$$S = \sqrt{P^2 + D^2} = \sqrt{(654)^2 + (540)^2} = 849 \text{ VA}.$$

Conclusion

Since the reactive power is zero, the apparent power has only real and distortive components. The distortive power arises from the current harmonics above the fundamental.

EXERCISE 1.19

Find the reactive and distortive powers of the circuit in Exercise 1.16.

Answer

$Q = 0$ and $D = 2.33$ kVAd.

Power Factor Correction in Nonlinear Circuits with Sinusoidal Sources

As stated earlier, the power factor of a nonlinear circuit with positive VAr is improved with the placement of shunt capacitance. The value of capacitance is determined by Equation 1.66, even though the load current is nonsinusoidal. The expression is still valid because the voltage across the capacitor is of single frequency; consequently, the current through the capacitor is of single frequency as well. The capacitor current is 180° out-of-phase with one of the two components of the fundamental current. The capacitor current thus cancels one of the components and brings the remaining component in phase with the source voltage. The components of the fundamental current are determined as follows: with Equations 1.72 and 1.77, the fundamental sine coefficient is expressed as

$$b_1 = I_1 \cos\phi_1. \tag{1.96}$$

Equation 1.96 suggests the geometry illustrated in Figure 1.11.

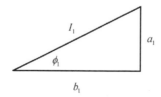

FIGURE 1.11   Geometry of the Fourier coefficients of the fundamental current.

Figure 1.11 implies the relationship for a_1 as

$$a_1 = I_1 \sin\phi_1. \tag{1.97}$$

Also implied by Figure 1.11 is an alternate relationship for the DPF:

$$\cos\phi_1 = \frac{b_1}{\sqrt{a_1^2 + b_1^2}}. \tag{1.98}$$

From Equations 1.92 and 1.97, an alternate formula for the reactive power is

$$Q = -\frac{a_1 V_m}{2}. \tag{1.99}$$

If the negative reactive power of the capacitor cancels the positive Q of the nonlinear load, it follows that $a_1 = 0$ for a value of capacitance expressed by Equation 1.66. The phasor relationships of the capacitor current and the fundamental current with its components are illustrated in Figure 1.12.

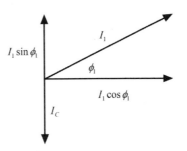

FIGURE 1.12 Phasor relationships of capacitor current and fundamental current.

If $a_1 = 0$, the displacement factor becomes

$$\cos\phi_1 = \frac{b_1}{\sqrt{a_1^2 + b_1^2}} = \frac{b_1}{\sqrt{b_1^2}} = 1.$$

If the displacement factor is unity, it follows from Equation 1.82 that the power factor and distortion factor are equal:

$$pf_C = DF_C. \tag{1.100}$$

In contrast to linear circuits, the power factor of a nonlinear circuit cannot be corrected to unity by the addition of shunt capacitance. The capacitor only cancels the reactive power; it has no effect on the distortive volt-amperes. With the shunt capacitor in the circuit and zero reactive power, the apparent power becomes

$$S_C = \sqrt{P^2 + D^2} \qquad (1.101)$$

and the improved power factor is

$$pf_C = \frac{P}{\sqrt{P^2 + D^2}}. \qquad (1.102)$$

The next example illustrates the process of power factor improvement in non-linear circuits.

EXAMPLE 1.20

Find the value of shunt capacitance to improve the power factor of a circuit that draws the current in Example 1.17 and is supplied by the source in Example 1.16. Find the real, reactive, and apparent powers and the power, displacement, and distortion factors with and without the capacitor in the circuit.

Solution

The phase angle between the source voltage and the current fundamental is

$$\phi_1 = \angle v - \angle i_1 = 0° - (-30°) = 30°.$$

The average, reactive, distortive, and apparent powers are computed as follows:

$$P = V_{RMS}\, I_{1,RMS} \cos\phi_1 = (120)(14.1)\cos(30°) = 1.47 \text{ kW.}$$

$$Q = V_{RMS}\, I_{1,RMS} \sin\phi_1 = (120)(14.1)\sin(30°) = 849 \text{ VAr.}$$

$$D = V_{RMS}\sqrt{I_{RMS}^2 - I_{1,RMS}^2} = 120\sqrt{(20.6)^2 - (14.1)^2} = 1.8 \text{ kVAd.}$$

$$S = \sqrt{P^2 + Q^2 + D^2} = \sqrt{(1.47)^2 + (0.85)^2 + (1.8)^2} = 2.47 \text{ kVA.}$$

The power, displacement, and distortion factors are calculated as follows:

$$pf = \frac{P}{S} = \frac{1.47}{2.47} = 0.59.$$

$$DPF = \cos\left(30°\right) = 0.87.$$

$$DF = \frac{14.1}{20.6} = 0.69.$$

The shunt capacitance required to improve the power factor is

$$C = \frac{Q_{L}}{\omega V_{RMS}^2} = \frac{849}{(120\pi)(120)^2} = 156 \ \mu F.$$

With this value of shunt capacitance, the reactive power is zero and the real and distortive powers are unchanged (on the assumption that the source impedance is negligible). The apparent power becomes

$$S_c = \sqrt{P^2 + D^2} = \sqrt{(1.47)^2 + (1.8)^2} = 2.32 \ \text{kVA},$$

and the new power factor is

$$pf_c = \frac{P}{\sqrt{P^2 + D^2}} = \frac{1.47}{2.32} = 0.63.$$

The new distortion factor is equal to the power factor, and the displacement factor is unity.

Conclusion

The shunt capacitor provides little improvement of the power factor in this example. The small improvement is due to the large VAd component of the apparent power over which the capacitor has no effect. The addition of shunt capacitance is practical only when there is a significant amount of VAr in the load.

EXERCISE 1.20

Find the value of shunt capacitance to improve the power factor of a circuit that draws the current in Exercise 1.17 and is supplied a 60 Hz source $v(\theta) = 220\sqrt{2} \cos\theta$ V with negligible impedance. Find the real, reactive, distortive, and apparent powers and the power, displacement, and distortion factors, with and without the shunt capacitor in the circuit.

Answers

$P = 751$ W

$Q = 201$ VAr

$D = 381$ VAd

$S = 866$ VA

$pf = 0.87$

$DPF = 0.97$

$DF = 0.90$

$C = 11\ \mu\text{F}$

$P_C = 751$ W

$Q_C = 0$

$D_C = 381$ VAd

$S_C = 842$ VA

$pf_C = 0.89$

$DPF_C = 1$

$DF_C = 0.89.$

Linear Circuits with Nonsinusoidal Sources

When an electrical network with linear circuit elements is supplied by a periodic, nonsinusoidal source, the current drawn from the source is also periodic and non-sinusoidal. The applied voltage, expressed in terms of the compact Fourier series, is

$$v(\theta) = V_0 + \sum_{n=1}^{\infty} V_n \cos(n\theta - \phi_{n,v}). \tag{1.103}$$

The current drawn by the network supplied by the voltage of Equation 1.103 is

$$i(\theta) = I_0 + \sum_{n=1}^{\infty} I_n \cos(n\theta - \phi_{n,i}). \tag{1.104}$$

The product of the instantaneous voltage and current in these Fourier series forms yields products of sinusoids of both like and unlike frequencies. As shown earlier, the orthogonal sinusoids contribute nothing to the average power. Only the products of like-frequency sinusoids result in nonzero integrals for the computation of average power. Therefore, for an element with instantaneous voltage and current expressed by Equations 1.103 and 1.104, the average power absorbed in the element is computed by

$$P = V_0 I_0 + \sum_{n=1}^{\infty} V_{n,RMS}\, I_{n,RMS} \cos(\phi_{n,v} - \phi_{n,i}). \tag{1.105}$$

The next example illustrates an application of Equation 1.105.

EXAMPLE 1.21

Find the power absorbed by a circuit element with voltage and current waveforms

$$v(\theta) = 120 + 40\cos(120\pi t - 30°) + 15\cos(240\pi t + 45°).$$

and

$$i(\theta) = 10 + 8\cos(240\pi t + 15°) + 3\cos(360\pi t - 60°).$$

Solution

The product of instantaneous voltage and current has like-frequency sinusoids at 0 and 120 Hz. Thus, the average power absorbed is

$$P = (120)(10) + \frac{(15)(8)}{2}\cos 30° = 1.25 \text{ kW}.$$

Conclusion

Both the voltage and current waveforms have a DC component and a harmonic at 120 Hz. Only these two terms contribute to the absorption of average power in the circuit element.

EXERCISE 1.21

Find the power absorbed in the element of Example 1.21 if the current harmonics are at the same frequencies as the voltage harmonics.

Answer

$P = 1.31$ kW.

The next example illustrates the power computations for a linear circuit supplied by a nonsinusoidal source.

EXAMPLE 1.22

A linear circuit supplied by a nonsinusoidal source is shown in Figure 1.13. Find the average power absorbed by the circuit if the source voltage is expressed as

$$v_s(t) = 100 + 50\cos(120\pi t) + 25\cos(240\pi t) + 15\cos(360\pi t).$$

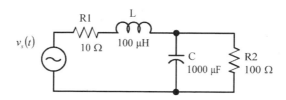

FIGURE 1.13   Linear circuit with nonsinusoidal source.

Solution

Since the reactive elements do not absorb average power, the power supplied by the source is equal to the sum of the powers dissipated in the two resistors. The power absorbed by the resistors is calculated with the RMS voltage or RMS current associated with each resistance and application of the DC power formulas as appropriate. The total impedance seen by the source, Z_T, is equal to the parallel combination of R_2 and jX_c plus the series combination of R_1 and jX_L. The impedance is computed at each harmonic frequency of the source voltage. The code in Listing 1.8 performs the computations. The program is available on the CD_ROM as Example1_22.m in Chapter 1\Examples and Listings.

ON THE CD

LISTING 1.8   Solution to Example 1.22

```
close all, clear all, clc
uH = 1E-6;
uF = 1E-6;
w = 2*pi*[1E-9 60 120 180];   %1E-9 prevents division by zero.
Vs = [100 [50 25 15]/sqrt(2)];
R1 = 10;
R2 = 100;
L = 100*uH;
C = 1000*uF;
XC = -1./(w*C);
XL = w*L;
Zp = R2*j*XC./(R2 + j*XC);    % R2 and C in parallel
Zs = R1 + j*XL;         % R1 and L in series
ZT = Zs + Zp;  % Total circuit impedance
Is = Vs./(abs(ZT));
Is_RMS = sqrt(sum(Is.^2));    % RMS source current
P1 = Is_RMS^2 * R1    % Power absorbed by R1
V2 = Vs.*abs(Zp./ZT); % Voltage across R2
V2_RMS = sqrt(sum(V2.^2));    % RMS voltage across R2
```

```
P2 = V2_RMS.^2/R2
phi = angle(ZT);
Ps = sum(Vs.*Is.*cos(phi))    % Also P1 + P2
```

Conclusion

The computed results of the code in Listing 1.8 are $P_1 = 166$ W, $P_2 = 84$ W, and $P_s = 250$ W. The magnitudes of the source current harmonics are computed from the magnitudes of the voltage harmonics divided by the magnitudes of the total impedance computed at each harmonic frequency. Since R_1 is in series with the source, the RMS source current determines the power dissipation in R_1. The voltage across R_2 is found by voltage division. The power delivered by the source is the sum of the average powers absorbed at each harmonic frequency.

EXERCISE 1.22

Compute the power delivered by the source of the circuit shown in Figure 1.14 with the same source voltage as in Example 1.22.

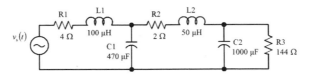

FIGURE 1.14   Circuit for Exercise 1.22.

Answer

$P_s = 250$ W.

Circuits with Nonsinusoidal Sources and DC Battery Loads

Many circuits supplied by a nonsinusoidal voltage source contain linear circuit elements and one or more DC sources. Such is the case of a battery charger in which the DC source (the battery) is the load. Battery-charging circuits typically supply a highly nonsinusoidal current to the battery; however, the voltage across the battery is essentially constant. (The battery voltage slowly increases over time as it is charged.) If the voltage across a circuit element is constant, the average power absorbed is

$$P = \frac{1}{T}\int_0^T V_b i(t)\,dt = V_b\left[\frac{1}{T}\int_0^T i(t)\,dt\right]. \tag{1.106}$$

The quantity in brackets in Equation 1.106 is, by definition, the average current. The average power absorbed by a battery, therefore, is

$$P_b = V_b I_b, \tag{1.107}$$

in which I_b is the average current through the battery. Equation 1.107 dictates that only the average value of current contributes to the average power absorbed by the battery. Consequently, only the DC component of current delivers charge to the battery. The current harmonics deliver charge to the battery during the first half-cycle of the harmonic waveform, only to return the charge back to the source during the second half-cycle. Consequently, the current harmonics only result in power dissipation in the series resistances of the battery and the circuit, with no net charge delivered.

EXAMPLE 1.23

Find the average power absorbed by the battery in the circuit shown in Figure 1.15, in which the source is the voltage of Example 1.18 with $V_m = 50$ V. How much charge has entered the battery after a 2-hour period?

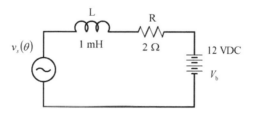

FIGURE 1.15 Battery charging circuit.

Solution

The average value of the battery current is due solely to the DC component of the source voltage. Since the inductor appears as a short circuit to the average voltage, the average current through the circuit is the average voltage across resistor R divided by the resistance:

$$I_b = \frac{\dfrac{2V_m}{\pi} - V_b}{R} = \frac{\dfrac{100}{\pi} - 12}{2} = 9.92 \text{ A}.$$

The power absorbed by the battery therefore is

$$P_b = V_b I_b = (12)(9.92) = 119 \text{ W}.$$

The charge absorbed by the battery is the product of the average current and charging time:

$$Q = I_b t = (9.92)(2)(60)(60) = 7.14 \times 10^4 \text{ C}.$$

Conclusion

The average current is readily calculated by the average voltage across the series resistance in the circuit. The value of the inductance is irrelevant to the average current, as it appears as a short circuit to the average voltage. The inductance does, however, limit the magnitudes of the current harmonics.

EXERCISE 1.23

Find the average power absorbed by the battery in the circuit shown in Figure 1.16, in which the source is the current waveform from Example 1.17. How much charge has entered the battery after 30 minutes?

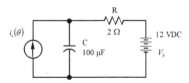

FIGURE 1.16 Battery charging circuit from Exercise 1.23.

Answers

$P_b = 1200$ W and $Q = 18000$ C.

CHAPTER SUMMARY

- Instantaneous power in an electrical circuit is the product of instantaneous voltage and instantaneous current.
- Instantaneous power represents the rate at which energy is expended.
- Ideal inductors and capacitors:
  - Store energy but do not dissipate energy.
  - Absorb zero average power when the voltage and current waveforms are periodic.
- RMS values of voltage and current are used to compute average power using the DC power formulas derived from Ohm's law.
- The definitions of apparent power, average power, and power factor are universal; they apply to all periodic voltage and current waveforms.
- A sinusoidal circuit has sinusoidal voltage and current waveforms that are all of the same frequency.
  - The apparent power has two components: average power and reactive power.
  - A lagging power factor refers to an inductive load; the current in the load lags the voltage, and the reactive power is positive.
  - A leading power factor refers to a capacitive load; the current in the load leads the voltage, and the reactive power is negative.
  - A lagging power factor can be corrected to unity with shunt capacitance.
  - A leading power factor cannot be corrected with capacitance.
- The apparent power of a circuit that draws nonsinusoidal current from a sinusoidal source has three components: average power, reactive power, and distortive power.
  - The reactive power is due to a phase displacement between the voltage and the fundamental component of current.
  - The distortive power is due to current harmonics above the fundamental frequency.
  - The displacement factor is the cosine of the angle between the source voltage and the fundamental component of current.
  - The distortion factor is the ratio of the RMS value of the fundamental component of current to the total RMS value of the current.
  - The power factor of a circuit with positive reactive power can be improved with shunt capacitance, but it cannot be corrected to unity because of the current harmonics.
- The average power absorbed by a circuit element with a constant voltage across it is the product of the constant voltage and the average current through the element.

THE MATLAB TOOLBOX

AVG

Function *avg.m* computes the average value of a vector.

Syntax

```
Xavg = avg(t, xt)
```

Vector **t** is the time interval over which the average value of **xt** is computed. Vectors **t** and **xt** must be the same length.

EXAMPLE

Code segment computes the average value of $x(t) = x(t+T) = 10t \sin(2\pi t)$.

```
t = linspace(0, 1, 1024);
vt = 10*t.*sin(2*pi*t);
V0 = avg(t, vt)
```

RMS

Function *rms.m* computes the RMS value of a vector.

Syntax

```
Xrms = rms(t, xt)
```

Vector **t** is the time interval over which the RMS value of **xt** is computed. Vectors **t** and **xt** must be the same length.

EXAMPLE

Code segment computes the RMS value of $x(t) = x(t+T) = 10t \sin(2\pi t)$.

```
t = linspace(0, 1, 1024);
vt = 10*t.*sin(2*pi*t);
Vrms = rms(t, vt)
```

FOURS

Function *fours.m* computes and plots the Fourier series expansion of a periodic function.

Syntax

```
fours(t, xt, N)
```

Vector **t** is one period of **xt**; N is the number of harmonics computed. Vectors **t** and **xt** must be the same length.

EXAMPLE

Code segment computes and plots the Fourier series expansion of

$$x(t+T) = \begin{cases} 1, 0 \leq t \leq \dfrac{1}{2} \\ 0, \dfrac{1}{2} < t \leq 1 \end{cases} :$$

```
N = 10;
t = linspace(0, 1, 1024);
xt = [ones(1, 512) zeros(1, 512)];
fours(t, xt, N).
```

SINPOW

Function *sinpow.m* computes the RMS current through a linear impedance supplied by a sinusoidal voltage source along with the apparent, average, real, and reactive powers and the power factor. If Q is positive, the PFC capacitor and the corrected load impedance are also computed.

Syntax

```
[Irms, S, P, Q, pf] = sinpow(f, Vs, Z)
```

Vs is the rectangular form of a sinusoidal voltage source of frequency *f* and RMS value *abs(Vs)*. *Z* is the load impedance in rectangular form computed at frequency *f*.

EXAMPLE

Code segment computes I_{RMS}, S, P, Q, pf, C, and Z_c of a load impedance $Z = 3 + j4\Omega$ supplied by a $110 + j50$ VAC voltage source at 60 Hz.

```
Vs = 110 + j*50;
f = 60;
ZL = 3 + j*4;
[Irms, S, P, Q, pf] = sinpow(f, Vs, ZL)
```

NONSIN

Function *nonsin.m* computes the apparent, real, reactive, and distortive powers and the power factor, displacement factor, and distortion factor. If Q is positive, the shunt capacitance, improved apparent power, and improved power factor are also computed.

Syntax

```
[S, P, Q, D, pf, DPF, DF] = nonsin(th, Ith, Vm, f)
```

Vector **th** is one radian period of **Ith**; **th** and **Ith** must be the same length. *Vm* is the peak voltage of a sinusoidal voltage source of frequency *f*.

EXAMPLE

Code segment computes S, P, Q, D, pf, DPF, and DF for a source current

$$i(\theta + 2\pi) = \theta \sin \theta$$

drawn from a 120 VAC source.

```
Vm = 120*sqrt(2);
th = linspace(0, 2*pi, 1024);
Ith = th.*sin(th);
[S, P, Q, D, pf, DPF, DF] = nonsin(th, Ith, Vm, f)
```

PROBLEMS

Problem 1

Air conditioning systems are rated in units of *tons of refrigeration*. The unit does not refer to weight, however; it is a unit of power related to the British thermal unit (Btu). One ton of refrigeration is equal to 3 1/3 Btu per second, and 1 Btu/s is equal to 1.055 kW. Find the cost to operate a 2.5-ton air conditioner for 6 hours a day for 30 days if electricity costs $0.12 per KWH.

Problem 2

The instantaneous voltage across and current through a device are, respectively,

$$v(\theta + \pi) = 100 \sin^2 \theta$$

and

$$i(\theta + \pi) = 10 \cos^2 \theta.$$

Plot the instantaneous voltage, current, and power. Compute the energy absorbed by the device in one source period. Compute the average power.

Problem 3

The voltage across a 1000 μF capacitor is

$$v(\theta + 2\pi) = 100 \theta^2 e^{-5\theta/\pi}.$$

Compute and plot the instantaneous power absorbed by the capacitor. Compute the average power absorbed by the capacitor.

Problem 4

The current through a 1 mH inductor is

$$i(\theta + \pi) = \theta^3 \sin \theta.$$

Compute and plot the instantaneous power absorbed by the inductor. Compute the average power absorbed by the inductor.

Problem 5

A load impedance $Z = 20 + j30\ \Omega$ is supplied by a source voltage $V_s = 440\angle 30°$ VAC. Compute the load current along with the apparent, real, and reactive powers, and the power factor. Compute the shunt capacitance to correct the power factor. How much current is drawn by the load after the power factor is corrected?

Problem 6

A 10 kW load with a power factor of 0.8 is supplied through a distribution line by a 460 VAC, 60 Hz source. The impedance of the line is $Z_{line} = 0.6 + j0.2\ \Omega$. The load voltage is 430 VAC. Compute the efficiency of the circuit. Determine the shunt capacitance to correct the power factor and recompute the efficiency. What is the source power factor after the load power factor has been corrected?

Problem 7

A nonlinear circuit draws the current

$$i(\theta + \pi) = \theta^4 \sin\theta$$

from a 120 VAC sinusoidal source. Compute the apparent, real, reactive, and distortive powers along with the power factor, displacement factor, and distortion factor. Can the power factor be improved with shunt capacitance? If so, compute the capacitance and the improved power factor.

Problem 8

Compute and plot two cycles of eight harmonics of the Fourier series expansion of the current waveform from Problem 7.

Problem 9

A 36 V battery draws the periodic current from Problem 7. Determine how much charge has entered the battery after a 4-hour period.

Problem 10

A nonlinear load is supplied by a source voltage expressed as

$$v_s(\theta) = 100 + 50\sin\left(\theta - 45°\right) + 25\cos\left(2\theta - 30°\right) + 10\sin\left(4\theta - 15°\right)$$

and draws current from the source expressed by

$$i_s(\theta) = 10 + 5\cos\left(\theta - 30°\right) + 5\sin\left(2\theta + 45°\right) + 2\sin\left(3\theta + 15°\right).$$

Determine the average power absorbed by the load and the load power factor.

Problem 11

The circuit of Figure 1.15 is supplied by the voltage source

$$v_s(\theta) = 40 + 5\sin\left(\theta + 45°\right) + 2\cos\left(2\theta + 30°\right) + \sin\left(3\theta + 15°\right).$$

Determine the expression for the source current. Compute the circuit efficiency.

Problem 12

The circuit of Figure 1.16 is supplied by the current source

$$v_s(\theta) = 5 + 3\sin\left(\theta + 30°\right) + 2\cos\left(2\theta - 15°\right) + \sin\left(4\theta - 15°\right).$$

Determine the expression for the capacitor voltage. Compute the circuit efficiency.

Part

II

Rectification

The electrical energy produced by utility companies is in the form of *alternating current* (AC) even though virtually all electronic systems require *direct current* (DC). The discrepancy arises for reasons of efficiency and economy: AC energy can be transformed to very high voltages that result in lower energy losses in the electrical distribution lines. The high voltage is then transformed again to a low AC voltage that is subsequently converted to DC by the rectifier circuit.

2 Diode Rectifier Circuits

In This Chapter

- Introduction
- Rectification
- Rectifier Circuits with Load Capacitance
- Chapter Summary
- Problems

INTRODUCTION

This chapter presents the theory and analysis of rectifier circuits. The method of analysis begins with a derivation of the instantaneous source current by the application of circuit laws. The analysis continues with the computation of the average load current, average load voltage, and average diode current. The average diode current is computed from the half-cycle average of the source current and determines the power dissipation in the device.

In the circuit analyses, the source voltage is assumed to be purely sinusoidal. The source current, however, is nonsinusoidal, and thus the power theory developed in Chapter 1 for nonlinear circuits is applicable. The analyses conclude with the numerical evaluations of RMS source current, the apparent, average, reactive, and distortive powers, and the power, displacement, and distortion factors. The results of the computations are followed by a discussion of possible power factor improvement. The study and analysis of rectifier circuits are essential because the rectifier is a key subcircuit in AC-sourced DC power supplies and battery chargers.

RECTIFICATION

The purpose of a rectifier circuit is to produce a DC voltage (VDC) and current from an AC source. There are two types of rectification processes: controlled rectification and uncontrolled rectification. Controlled rectification is accomplished by silicon-controlled rectifiers (SCRs), also known as thyristors; uncontrolled rectification is performed by diodes.

Diodes

The diode is a two-terminal device that conducts current in only one direction. The two terminals are the anode (positive) and the cathode (negative), as illustrated in Figure 2.1.

anode   cathode

FIGURE 2.1   A semiconductor diode.

When the voltage across the diode is positive from anode to cathode, the diode is *forward-biased* and conducts current in the direction of anode to cathode. If the voltage is positive from cathode to anode, the device is *reverse-biased* and no current flows. The voltage across the diode when forward-biased is called the *forward voltage, V_f.* In rectifier circuits this voltage is typically much smaller than the voltages of the source and the load. Unless power dissipation in the diode is under consideration, the forward voltage is usually neglected in the analysis of rectifier circuits.

Since the diode only conducts when forward-biased, the device is effectively approximated as a simple on/off switch. When used in a circuit with an AC source, the diode is alternately forward-biased and reverse-biased by the alternating polarity of the source voltage.

Half-Wave Rectifier with Resistive Load

The simplest rectifier circuit is the half-wave (HW) rectifier with resistive load. The HW rectifier has a single diode as shown in Figure 2.2.

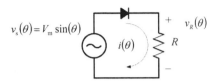

FIGURE 2.2 Half-wave rectifier with resistive load.

The source voltage is described as

$$v_s(\theta) = V_m \sin\theta \tag{2.1}$$

and is typically the secondary voltage of a supply transformer. When v_s is positive, the diode is forward-biased and the source voltage appears directly across the load. When v_s is negative, the diode is reverse-biased and the load is effectively disconnected from the source. The switching function of the diode is illustrated in Figure 2.3.

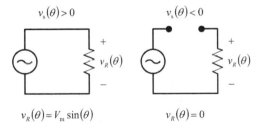

FIGURE 2.3 Equivalent circuits of the HW rectifier.

No current flows through the load during the negative half-cycle of the source; consequently, the load voltage is zero. The load voltage and current waveforms are illustrated in Figure 2.4.

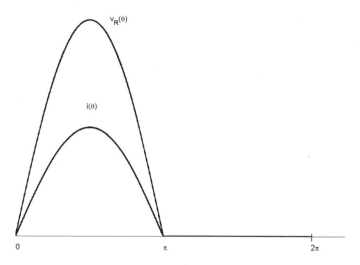

FIGURE 2.4 Load voltage and current waveforms of the HW rectifier.

Average Load Current

The load voltage and current waveforms are expressed as follows:

$$v_R(\theta) = \begin{cases} V_m \sin\theta, & 0 \le \theta \le \pi \\ 0, & \pi < \theta \le 2\pi \end{cases}.$$ (2.2)

$$i(\theta) = \begin{cases} \dfrac{V_m}{R} \sin\theta, & 0 \le \theta \le \pi \\ 0, & \pi < \theta \le 2\pi \end{cases}.$$ (2.3)

As stated in the introduction, the purpose of rectification is to produce a DC voltage and current from an AC source. The average output voltage is determined from the average value integral presented in Chapter 1:

$$V_o = \frac{1}{2\pi} \int_0^\pi V_m \sin\theta \, d\theta.$$

$$V_o = -\frac{V_m}{2\pi} \cos\theta \Big|_0^\pi.$$

$$V_o = \frac{V_m}{\pi}.$$ (2.4)

Equation 2.4 reveals that the output voltage of the HW circuit is roughly one-third of the peak source voltage. The average load current is determined with Equation 2.3, or more simply, is the average load voltage divided by the resistance:

$$I_o = \frac{V_m}{\pi R}. \tag{2.5}$$

Since the diode and load resistor are in series, the average diode current is the same as the average load current. The power dissipation in the diode, therefore, is

$$P_{DIO} = \frac{V_f V_m}{\pi R}. \tag{2.6}$$

Power Computations

Application of the RMS value definition to Equation 2.3 yields the RMS source (and load) current:

$$I_{RMS} = \sqrt{\frac{1}{2\pi} \int_0^\pi \frac{V_m^2}{R^2} \sin^2 \theta \, d\theta}. \tag{2.7}$$

Evaluation of Equation 2.7 results in an RMS value of

$$I_{RMS} = \frac{V_m}{2R}. \tag{2.8}$$

With Equation 2.8, the power output to the load is

$$P_o = \frac{V_m^2}{4R}. \tag{2.9}$$

The circuit efficiency is the ratio of the output power absorbed to the sum of the output power and the power dissipated in the diode:

$$\eta = \frac{P_o}{P_o + P_{DIO}}. \tag{2.10}$$

Substitution of Equations 2.6 and 2.9 into Equation 2.10 yields the efficiency expression

$$\eta = \frac{1}{1 + \dfrac{4}{\pi} \dfrac{V_f}{V_m}}. \tag{2.11}$$

The product of RMS source voltage and RMS current is the apparent volt-amperes,

$$S = \frac{V_m^2}{2\sqrt{2}R},$$ (2.12)

and the power factor is the ratio of Equation 2.9 to Equation 2.12:

$$pf = \frac{1}{\sqrt{2}}.$$ (2.13)

Equation 2.13 reveals that the power factor is not dependent upon any circuit parameter; it is simply 0.707 for the HW rectifier with resistive load. Although the power factor is rather poor, the HW rectifier is the most economical rectifier circuit because only one diode is required.

Reactive and Distortive Powers

The fundamental cosine coefficient of the source current of the HW rectifier is determined as follows:

$$a_1 = \frac{1}{2\pi}\int_0^\pi \frac{V_m}{R}\sin\theta\cos\theta\,d\theta.$$

$$a_1 = \frac{V_m}{4\pi R}\int_0^\pi \sin 2\theta\,d\theta = 0.$$

The reactive power is thus zero, and the poor power factor described by Equation 2.13 is due entirely to current harmonics. No power factor improvement with shunt capacitance is possible. Furthermore, with $a_1 = 0$, the displacement factor is unity and the power and distortion factors are equal. In this case, the distortive volt-amperes are determined by Equation 2.9, the same expression as the average power.

EXAMPLE 2.1

A certain HW rectifier with a 10 Ω load resistance is supplied by a 120 VAC source. Compute the average load voltage and current, the RMS source current, and the average and apparent powers of the source. Compute the diode power dissipation and the efficiency if the forward voltage is 1.7 V.

Solution

The computations are performed by the MATLAB code in Listing 2.1. The program is available on the CD-ROM as file Example2_1.m in folder Chapter 2\Examples and Listings.

LISTING 2.1   Computations for Example 2.1

```
close all, clear all, clc
Vs = 120;
R = 10;
Vf = 1.7;
Vm = 120*sqrt(2);
Vo = Vm/pi
Io = Vo/R
Is = Vm/(2*R)
P = Vs^2/(2*R)
S = Vs*Is
Pdio = Vf*Io
eta = 1/(1 + 4/pi*Vf/Vm)
```

The computations produced by the code in Listing 2.1 are

$V_o = 54.0$ VDC

$I_o = 5.40$ A

$I_s = 8.49$ A

$P = 720$ W

$S = 1.02$ kVA

$P_{DIO} = 9.18$ W

$\eta = 98.7\%$.

Conclusion

The results reveal a significantly larger RMS source current than average load current—a typical indicator of a poor power factor. The power factor and distortive volt-amperes are readily verified by the results of the program: the power factor is 720/1.02E3 = 0.71 and the distortive power is $\sqrt{(1.02\text{E}3)^2 - (720)^2} = 720$ VAd.

EXERCISE 2.1

Find the peak source voltage required for a half-wave rectifier to produce 100 VDC across a 20 Ω resistance. What average power is absorbed by the load?

Answers

$V_m = 314$ V and $P = 1.23$ kW.

Exact versus Approximate Analyses

If the assumption of negligible diode voltage is not valid, the source voltage is diminished by the forward voltage,

$$V_s(\theta) = V_m \sin\theta - V_f, \tag{2.14}$$

and the diode does not conduct current until the source voltage exceeds the forward voltage drop. The point after which the diode becomes forward-biased is the solution of Equation 2.14 for θ when the source voltage is zero:

$$\theta_1 = \arcsin\left(\frac{V_f}{V_m}\right). \tag{2.15}$$

By symmetry, the point after which the diode becomes reverse-biased is

$$\theta_2 = \pi - \theta_1. \tag{2.16}$$

The voltage and current waveforms of the HW rectifier with significant forward diode voltage are plotted in Figure 2.5.

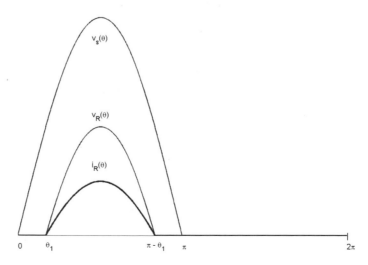

FIGURE 2.5   Waveforms of the HW rectifier with significant V_f.

If the forward voltage were truly zero, the efficiency would be 100%. The efficiency expression in Equation 2.10 resulted not from the assumption of negligible V_f but from the application of the conservation of energy principle: net energy in equals net energy out; nonetheless, the equation is an approximation because V_f was omitted from the source voltage expression.

There are thus two approaches to relate the efficiency and the forward voltage: (a) include V_f in the instantaneous load voltage and current expressions or (b) apply an alternate approximation. The first approach results in a set of equations that are much more algebraically complicated than those based on negligible V_f. The second approach directly yields an approximate efficiency expression. An alternate approximation is to assume that the peak source voltage alone is diminished by the forward voltage:

$$V_s(\theta) \approx \left(V_m - V_f \right) \sin\theta. \tag{2.17}$$

A comparison of Equation 2.17 with Equation 2.1 shows that the peak voltage is simply replaced by $V_m - V_f$. All the previous derivations are thus modified in the same manner. Application of the approximation to Equation 2.11 results in the efficiency expression

$$\eta = \frac{1}{1 + \dfrac{4}{\pi} \dfrac{V_m}{V_m - V_f}}. \tag{2.18}$$

Example 2.2 illustrates and compares the exact and approximate approaches to a circuit analysis problem and efficiency study. The example demonstrates that a viable approximation yields solutions that are widely applicable and much more tractable than their exact counterparts.

EXAMPLE 2.2

Derive the exact efficiency expression based on Equations 2.14 to 2.16. Express the exact efficiency and the approximate efficiency as functions of the ratio of V_f to V_m. Determine a range of suitable applicability of Equations 2.11 and 2.18.

Solution

Based on Equations 2.14 to 2.16, the instantaneous load voltage and current become

$$v_R(\theta) = \begin{cases} V_m \sin\theta - V_f, & \theta_1 \le \theta \le \pi - \theta_1 \\ 0, & \pi - \theta_1 < \theta \le 2\pi + \theta_1 \end{cases}. \tag{2.19}$$

$$i(\theta) = \begin{cases} \dfrac{V_m}{R}\sin\theta - \dfrac{V_f}{R}, & \theta_1 \le \theta \le \pi - \theta_1 \\ 0, & \pi - \theta_1 < \theta \le 2\pi + \theta_1 \end{cases}. \tag{2.20}$$

The average value of load voltage expressed by Equation 2.19 is

$$V_o = \frac{V_m}{\pi}\cos\theta_1 - \left(\frac{1}{2} - \frac{\theta_1}{\pi}\right)V_f. \tag{2.21}$$

If V_f is zero, then θ_1 is also zero and Equation 2.21 reverts to Equation 2.4. The RMS value of the load current expressed by Equation 2.20 is

$$I_{\text{RMS}} = \frac{V_m}{2R}\sqrt{\left[\left(1 - \frac{2\theta_1}{\pi}\right)\left(1 + 2\gamma^2\right) + \frac{1}{\pi}\sin 2\theta_1 - \frac{8\gamma}{\pi}\cos\theta_1\right]}, \tag{2.22}$$

in which γ (gamma) is the ratio of the forward diode voltage to the peak source voltage:

$$\gamma = \frac{V_f}{V_m}. \tag{2.23}$$

If V_f is zero, then γ is zero and Equation 2.22 reverts to Equation 2.8. With Equations 2.21 and 2.23, the diode dissipation is expressed as

$$P_{\text{DIO}} = \frac{V_m^2}{R}\left[\frac{\gamma}{\pi}\cos\theta_1 - \gamma^2\left(\frac{1}{2} - \frac{\theta_1}{\pi}\right)\right], \tag{2.24}$$

and with Equations 2.22 and 2.23, the power absorbed in the load is

$$P_o = \frac{V_m^2}{R}\left[\left(\frac{1}{2} - \frac{\theta_1}{\pi}\right)\left(\frac{1}{2} + \gamma^2\right) + \frac{1}{4\pi}\sin 2\theta_1 - \frac{2\gamma}{\pi}\cos\theta_1\right]. \tag{2.25}$$

Substitution of Equations 2.24 and 2.25 into Equation 2.10 yields the efficiency expression

$$\eta = \frac{\left(\pi - 2\theta_1\right)\left(1 + 2\gamma^2\right) + \sin 2\theta_1 - 8\gamma\cos\theta_1}{\pi - 2\theta_1 + \sin 2\theta_1 - 4\gamma\cos\theta_1}. \tag{2.26}$$

In terms of the gamma ratio, Equations 2.11 and 2.18 become, respectively,

$$\eta = \frac{1}{1 + \dfrac{4}{\pi}\gamma} \tag{2.27}$$

and

$$\eta = \frac{1}{1 + \dfrac{4}{\pi}\dfrac{\gamma}{1-\gamma}}. \tag{2.28}$$

Equations 2.26 to 2.28 are plotted as a function of the gamma ratio in Figure 2.6.

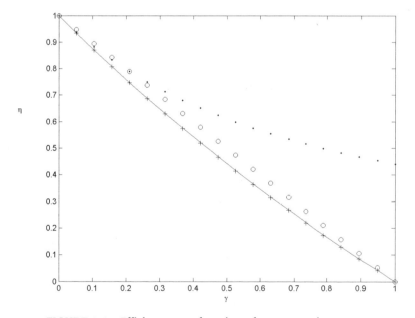

FIGURE 2.6   Efficiency as a function of gamma ratio.

Conclusion

The plots reveal that the approximation of Equation 2.28 is a virtually perfect estimation of the exact efficiency over the entire range of gamma. The approximation of Equation 2.27 is accurate only for gamma ratios less than 10%. Although highly inaccurate over the whole range of gamma, Equation 2.27 does provide a convenient vehicle by which to obtain the accurate approximation of Equation 2.28 by the replacement of V_m by $V_m - V_f$. The example clearly demonstrates that the use of simplifying assumptions is a worthwhile approach to analysis.

EXERCISE 2.2

Deduce another approximate expression for efficiency from the exact plot in Figure 2.6. Comment on the viability of the approximation.

Answer

$$\eta = 1 - \gamma. \tag{2.29}$$

The approximation is more accurate than Equation 2.27 but overestimates the efficiency. Equation 2.29 is also plotted in Figure 2.6.

Full-Wave Rectifier with Resistive Load

As illustrated earlier, the HW rectifier is disconnected from the source during the negative half-cycle. If the negative half of the source is rectified as well, the result is full-wave (FW) rectification. The FW rectifier with a resistive load is shown in Figure 2.7.

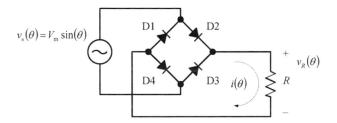

FIGURE 2.7   Full-wave rectifier with resistive load.

When the source voltage is positive, diodes D2 and D4 are forward-biased; D1 and D3 are reverse-biased. When the source voltage is negative, D2 and D4 are off, while D1 and D3 are on. The equivalent circuits in accordance with source polarity are shown in Figure 2.8.

As shown in Figure 2.8, the load current is unidirectional; hence, the load voltage polarity is always the same. The load voltage and load current waveforms are illustrated in Figure 2.9.

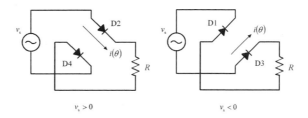

FIGURE 2.8   Equivalent circuits of the FW rectifier.

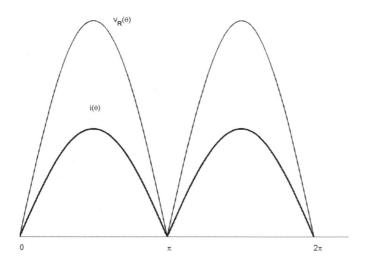

FIGURE 2.9   Load voltage and current waveforms of the
FW rectifier with resistive load.

The load voltage and current waveforms of Figure 2.9 are expressed as follows:

$$v_R(\theta) = V_m \sin\theta,\ 0 \le \theta \le \pi. \tag{2.30}$$

$$i(\theta) = \frac{V_m}{R} \sin\theta,\ 0 \le \theta \le \pi. \tag{2.31}$$

Average Load Voltage and Current

The period of the load voltage for the FW rectifier is π; thus, the average output
voltage of the FW rectifier is twice that of the HW rectifier:

$$V_o = \frac{2V_m}{\pi}. \tag{2.32}$$

The increase in the DC component comes at the expense of three additional diodes. The average load current is also twice that of the HW rectifier:

$$I_o = \frac{2V_m}{\pi R}. \tag{2.33}$$

The average diode current of the FW rectifier is the same as that of the HW circuit because only one pair of diodes is forward-biased during one cycle of the source voltage.

Power Computations

It was shown in Chapter 1 that the RMS value of a full-wave rectified sinusoid is the same as that of the sinusoid itself. The RMS load (and source) current, therefore, is

$$I_{RMS} = \frac{V_m}{\sqrt{2}R}, \tag{2.34}$$

and the power absorbed by the resistance is

$$P_o = \frac{V_m^2}{2R}. \tag{2.35}$$

The power output of the FW rectifier is thus twice that of the HW rectifier. Unlike the HW rectifier, the source current of the FW circuit *is* sinusoidal. Since both the voltage and current are sinusoidal and the load is purely resistive, the power factor is unity and the average and apparent powers are the same. The reactive power is zero because the source voltage and current are in phase, and the distortive power is zero because the source current has no harmonics above the fundamental. The power dissipated in the diodes is, as the HW circuit, determined by Equation 2.6 because each diode conducts for only half of the source voltage period.

Because of the additional diodes, the efficiency of the FW circuit is computed as

$$\eta = \frac{P_o}{P_o + 4P_{DIO}}. \tag{2.36}$$

Substitution of Equations 2.6 and 2.35 into Equation 2.36 results in

$$\eta = \frac{1}{1 + \dfrac{8}{\pi} \dfrac{V_f}{V_m}}. \tag{2.37}$$

As with Equation 2.27, the accuracy of the efficiency expression is much improved when the peak source voltage is diminished by the diode forward voltage. In the case of the FW rectifier, the source is diminished by two forward voltages:

$$\eta = \frac{1}{1 + \dfrac{8}{\pi} \dfrac{V_f}{V_m - 2V_f}}. \tag{2.38}$$

Expressed in terms of the gamma ratio of Equation 2.23, the efficiency of the FW rectifier is

$$\eta = \frac{1}{1 + \dfrac{8}{\pi} \dfrac{\gamma}{1 - 2\gamma}}. \tag{2.39}$$

EXAMPLE 2.3

Repeat Example 2.1 for the case of the FW rectifier.

Solution

The computations are performed by the MATLAB code in Listing 2.2. The code is available in the file Example2_3.m under folder Chapter 2\Examples and Listings.

LISTING 2.2 Computations for Example 2.3

```
close all, clear all, clc
Vs = 120;
Vf = 1.7;
R = 10;
Vm = Vs*sqrt(2);
y = Vf/Vm;
Vo = 2*Vm/pi
Io = Vo/R
Is = Vs/R
Po = Vs^2/R
```

```
S = Vs*Is
Pdio = Vm*Vf/(pi*R)
eta = 1/(1 + 8/pi*y/(1 - y))
```

The results produced by the code in Listing 2.2 are

$V_o = 108$ VDC

$I_o = 10.8$ A

$I_s = 12$ A

$P = 1.44$ kW

$S = 1.44$ kVA

$P_{DIO} = 9.18$ W

$\eta = 97.5\%$.

Conclusion

As stated earlier, the average output voltage and current and the output power of the FW rectifier are twice those of the HW rectifier. The diode dissipation is the same as the HW rectifier; however, the efficiency of the FW circuit is slightly less because of the dissipations in the three additional diodes. The apparent power is the same as the average power because of the unity power factor.

EXERCISE 2.3

Find the peak source voltage required for a full-wave rectifier to produce 100 VDC across a 20 Ω resistance. What average power is absorbed by the load?

Answers

$V_m = 157$ V and $P = 617$ W.

Inductive Loads

It is rare that an electrical load is modeled as a pure resistance. More commonly, there is significant inductance in series with the resistance. Furthermore, the source that supplies the rectifier circuit is nearly always the secondary winding of a transformer that has resistance and inductance itself. These resistive and inductive (R-L) elements are lumped together to facilitate the analysis of a circuit with an inductive load.

HW Rectifier with R-L Load

The HW rectifier with series resistance and inductance is shown in Figure 2.10.

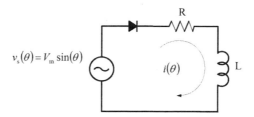

FIGURE 2.10   HW rectifier with R-L load.

Unlike the purely resistive case, the series current in the inductive circuit is described by the solution of a differential equation. Application of Kirchhoff's voltage law around the circuit loop in Figure 2.10 results in:

$$V_m \sin \omega t = iR + L\frac{di}{dt}. \tag{2.40}$$

Substitution of $\theta = \omega t$ transforms Equation 2.40 into

$$\frac{di}{d\theta} + \frac{R}{\omega L}i = \frac{V_m}{\omega L}\sin\theta. \tag{2.41}$$

Let a constant q be defined as

$$q = \frac{\omega L}{R}. \tag{2.42}$$

Equation 2.42 represents the *quality factor* of the inductance. In terms of the quality factor, the differential equation is expressed as

$$\frac{di}{d\theta} + \frac{i}{q} = \frac{V_m}{qR}\sin\theta. \tag{2.43}$$

The solution to Equation 2.43 is the sum of a forced response and a natural response:

$$i(\theta) = i_n(\theta) + i_f(\theta). \tag{2.44}$$

Since Equation 2.43 is a first-order differential equation, the natural response is

$$i(\theta) = A_0 e^{-\theta/q}, \tag{2.45}$$

in which the constant A_0 is determined from the initial conditions once the complete response has been formulated. The forcing function in Equation 2.43 is sinusoidal; consequently, the forced response is also a sinusoid:

$$i_f(\theta) = A_1 \cos\theta + A_2 \sin\theta. \tag{2.46}$$

Constants A_1 and A_2 are determined upon substitution of the forced response and its derivative into the differential equation:

$$-A_1 \sin\theta + A_2 \cos\theta + \frac{A_1}{q} \cos\theta + \frac{A_2}{q} \sin\theta = \frac{V_m}{qR} \sin\theta. \tag{2.47}$$

The coefficients of cosine on each side of Equation 2.47 produce the equation

$$A_2 + \frac{A_1}{q} = 0, \tag{2.48}$$

and the sine coefficients yield

$$-A_1 + \frac{A_2}{q} = \frac{V_m}{qR}. \tag{2.49}$$

Equations 2.48 and 2.49 form a system of equations for which the solutions are

$$A_1 = -\frac{qV_m}{R(1+q^2)}. \tag{2.50}$$

$$A_2 = \frac{V_m}{R(1+q^2)}. \tag{2.51}$$

Substitution of Equations 2.50 and 2.51 into Equation 2.46 yields the forced response

$$i_f(\theta) = -\frac{qV_m}{R(1+q^2)} \cos\theta + \frac{V_m}{R(1+q^2)} \sin\theta. \tag{2.52}$$

With Equation 1.10, the trigonometric identity from Chapter 1, the forced response is expressed as

$$i_f(\theta) = \frac{V_m}{R\sqrt{1+q^2}} \sin(\theta - \phi), \tag{2.53}$$

in which

$$\phi = \arctan(q).$$

(2.54)

Equation 2.54 suggests the geometric relationship illustrated in Figure 2.11.

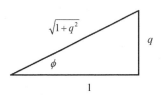

FIGURE 2.11   Geometric relationship between f and q.

From Figure 2.11, an additional relationship between ϕ and q is deduced:

$$\cos\phi = \frac{1}{\sqrt{1+q^2}}.$$

(2.55)

With Equations 2.45, 2.53, and 2.55, the complete response is formulated as

$$i(\theta) = A_0 e^{-\theta/q} + \frac{V_m}{R}\cos\phi\sin(\theta-\phi).$$

(2.56)

In the HW rectifier circuit, the current becomes zero before the beginning of the next cycle of the source voltage. Hence, the initial condition for Equation 2.56 is

$$i(0) = 0.$$

(2.57)

Substitution of the initial condition into Equation 2.56 yields the solution for A_0:

$$A_0 = \frac{V_m}{R}\cos\phi\sin\phi.$$

(2.58)

Substitution of Equation 2.58 into Equation 2.56 yields the final form of the instantaneous current of the HW rectifier with R-L load:

$$i(\theta) = \frac{V_m}{R}\cos\phi\left[\sin(\phi)e^{-\theta/q} + \sin(\theta-\phi)\right].$$

(2.59)

The instantaneous current relative to the HW rectified source voltage is illustrated in Figure 2.12.

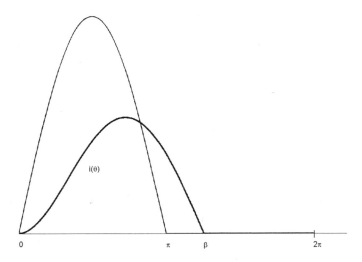

FIGURE 2.12 Instantaneous current of the HW rectifier with R-L load.

As shown in Figure 2.12, the load current becomes zero at $\theta = \beta$. Substitution of β into Equation 2.59 results in a relationship between β, ϕ, and q:

$$\sin(\phi)e^{-\beta/q} + \sin(\beta - \phi) = 0. \tag{2.60}$$

It is not possible to solve Equation 2.60 algebraically for β; it is a *transcendental equation* that must be solved numerically.

EXAMPLE 2.4

Write a MATLAB function that will return the value of β, given the value of q.

Solution

The solution is provided by the code in Listing 2.3. The function is available as file betasolv.m in folder Chapter 2\Toolbox.

LISTING 2.3 Function to Solve Equation 2.60 for β

```
function beta = betasolv(q)
phi = atan(q);
```

```
LHS = 1;
beta = pi;
inc = pi/2;
while(abs(LHS) > 1E-6),
    LHS = sin(phi)*exp(-beta/q) + sin(beta - phi);
    if LHS > 0,
        beta = beta + inc;
    else
        beta = beta - inc;
        inc = inc/2;
        beta = beta + inc;
    end
end
```

Conclusion

Angle β lies in the interval $\pi < \beta < 2\pi$; hence, the MATLAB function begins with a trial solution at $\beta = \pi$. If the value of the lefthand side (LHS) of Equation 2.60 is positive, then β is incremented by $\pi/2$, after which the LHS is computed with the new value of β. If the LHS is negative, β is decremented by $\pi/2$ and then incremented by $\pi/4$; the LHS is then evaluated once again. The process repeats as the increment is divided by two each time there is a sign change in the LHS. The process terminates when the LHS has a value less than 1E-6.

EXERCISE 2.4

Write a MATLAB program that uses the *betasolv* function to compute and plot the instantaneous source current as illustrated in Figure 2.12 with $V_m / R = q = 1$.

Average Load Current

The average load current is determined upon application of the average value integral to the instantaneous load current:

$$I_o = \frac{V_m}{2\pi R}\cos\phi \int_\alpha^\beta \left[\sin(\phi)e^{-\theta/q} + \sin(\theta-\phi)\right] d\theta. \tag{2.61}$$

Evaluation of the integrals in Equation 2.61 yields the average current expression

$$I_o = \frac{V_m}{2\pi R}\left[\sin^2\phi\left(1-e^{-\beta/q}\right)+\cos^2\phi-\cos\phi\cos(\beta-\phi)\right]. \tag{2.62}$$

Low *q* Approximate Solution

It is reasonable to approximate Equations 2.60 and 2.62 if the ratio of β to q is greater than 5. The approximation is valid in low q circuits—those with relatively small inductance and large resistance. In this case, the exponential terms in Equations 2.60 and 2.62 are essentially zero because $e^{-5} \approx 0$. If the approximation is valid, Equation 2.60 simplifies to

$$\sin(\beta - \phi) = 0, \tag{2.63}$$

and Equation 2.62 reduces to

$$I_o = \frac{V_m}{2\pi R}\left[1 - \cos\phi\cos(\beta - \phi)\right]. \tag{2.64}$$

The solution to Equation 2.63 that applies to the low q circuit is

$$\beta = \phi + \pi. \tag{2.65}$$

Substitution of Equation 2.65 into Equation 2.64 yields the approximate average current expression

$$I_o = \frac{V_m}{2\pi R}\left[1 + \cos\phi\right]. \tag{2.66}$$

Example 2.5 compares the approximate and exact solutions for the average current.

EXAMPLE 2.5

Find the average load current of a half-wave rectifier with R-L load if R = 5 Ω and (a) L = 10 mH and (b) L = 100 mH. The source voltage is 120 VAC at 60 Hz. Determine if the approximation is valid for each case of inductance.

Solution

Part (a): L = 10 mH. If the approximation is valid, then

$$\frac{\beta}{q} = \frac{\phi + \pi}{q} \geq 5.$$

From the source frequency and the component values, the quality factor is determined to be

$$q = \frac{\omega L}{R} = \frac{(120\pi)(0.01)}{5} = 0.75,$$

and the angle ϕ is

$$\phi = \arctan(q) = \arctan(0.75) = 0.65.$$

With these values of q and ϕ, the ratio of β to q is

$$\frac{\phi + \pi}{q} = \frac{0.65 + \pi}{0.75} = 5.02.$$

The approximation is thus valid, and the approximate average load current is

$$I_o = \frac{V_m}{2\pi R}\left[1 + \cos\phi\right] = \frac{120\sqrt{2}}{(2\pi)(5)}[1 + \cos(0.65)] = 9.72 \text{ A}.$$

The solution from Equations 2.60 and 2.62 is 9.70 amperes.

Part (b): L = 100 mH. The ratio of β to q in this case is $(\phi + \pi)/q = 0.61$. The approximation is therefore not applicable. The solution for β from Equation 2.60 is 5.11 radians, and the average load current computed from Equation 2.62 is 3.32 A.

Conclusion

The relatively large inductance of 100 mH substantially limits the average load current.

EXERCISE 2.5

A 240 VAC, 60 Hz source supplies a half-wave rectifier with R-L load. The load resistance is 10 Ω. On the assumption that the approximation is valid, find the value of L that produces an average load current of 10 A. Is the assumption valid?

Answer

$L = 16.4$ mH; the approximation is valid.

Power Computations and Efficiency

The diode dissipation is the product of the forward voltage and the average current:

$$P_{\text{DIO}} = \frac{V_f V_m}{2\pi R} \left[\sin^2 \phi \left(1 - e^{-\beta/q} \right) + \cos^2 \phi - \cos \phi \cos(\beta - \phi) \right]. \tag{2.67}$$

Since there is only one diode, the circuit efficiency is computed with Equation 2.10.

Although it is a straightforward process to derive the expressions for RMS source current, the apparent power and its three components, and the power factor and its two components, the equations that result are of such algebraic complexity as to limit their utility. In lieu of derivations, a numerical approach is preferable to make the power computations more tractable. The next example illustrates the numerical integration approach to power calculations.

EXAMPLE 2.6

Use MATLAB to perform the power computations for a half-wave rectifier with R-L load with circuit parameters $V_s = 120$ VAC, $f = 60$ Hz, $R = 25\ \Omega$, $L = 50$ mH, and $V_f = 1.2$ V. Is it possible to improve the circuit power factor with shunt capacitance?

Solution

ON THE CD

The solution is given by the code in Listing 2.4. The program is available as Example2_6.m under Chapter 2\Examples and Listings.

LISTING 2.4   Power Computations for the Circuit in Example 2.6

```
close all, clear all, clc
Vs = 120;
Vm = 120*sqrt(2);
f = 60;
T = 2*pi;
w = 2*pi*f;
R = 25;
L = 0.05;
q = w*L/R;
phi = atan(q);
beta = betasolv(q);
theta = linspace(0, beta, 1000);
Ith = Vm/R*cos(phi)*(sin(phi)*exp(-theta/q) + sin(theta - phi));
Is = sqrt(1/T*trapz(theta, Ith.^2));
```

```
a1 = 2/T*trapz(theta, Ith.*cos(theta));
b1 = 2/T*trapz(theta, Ith.*sin(theta));
Is1 = sqrt((a1^2 + b1^2)/2);
S = Vs*Is
P = Vm*b1/2
Q = -Vm*a1/2
D = Vs*sqrt(Is^2 - Is1^2)
pf = P/S
ph1 = atan(a1/b1);
DPF = cos(ph1)
DF = Is1/Is
```

Conclusion

The computations produced by the code in Listing 2.4 are

$I_{rms} = 2.90$ A

$S = 347$ VA

$P = 210$ W

$P_{DIO} = 2.33$ W

$Q = 124$ VAr

$D = 248$ VAd

$pf = 0.60$

$DPF = 0.86$

$DF = 0.70$

$\eta = 98.9\%$.

The RMS current and fundamental sine and cosine coefficients are quite easily computed with *trapz*. Since the reactive volt-amperes are positive, the power factor can be improved with shunt capacitance; however, the improvement is slight because of the significant amount of VAd. The efficiency is high because of the large difference between the peak source voltage and the diode forward voltage. In spite of the high efficiency, the diode dissipation is significant and the device may require thermal management.

EXERCISE 2.6

Compute the value of shunt capacitance that corrects the displacement factor in Example 2.6 to unity. Compute the improved power factor.

Answers

$C = 23\ \mu\text{F}$ and $pf = 0.65$.

Full-Wave Rectifier with R-L Load

The FW rectifier circuit with R-L load is shown in Figure 2.13.

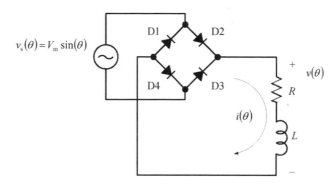

FIGURE 2.13   FW rectifier with R-L load.

The differential equation and general solution for the current of the FW circuit are the same as those of the HW circuit; however, *boundary* conditions apply rather than initial conditions. Because of the periodic and continuous nature of the current in the FW circuit, the boundary conditions are

$$i(0) = i(\pi). \tag{2.68}$$

Application of the boundary conditions to Equation 2.56 yields the relationship

$$A_0 + \frac{V_m}{R}\cos\phi\sin(-\phi) = A_0\,e^{-\pi/q} + \frac{V_m}{R}\cos\phi\sin(\pi-\phi). \tag{2.69}$$

The solution of Equation 2.69 for A_0 is

$$A_0 = \frac{2V_m}{R(1-e^{-\pi/q})}\cos\phi\sin\phi. \tag{2.70}$$

Substitution of Equation 2.70 into Equation 2.56 yields the complete response for the load current of the FW rectifier circuit:

$$i(\theta) = \frac{V_m}{R}\left[\frac{\sin 2\phi}{(1-e^{-\pi/q})}e^{-\theta/q} + \cos\phi\sin(\theta-\phi)\right]. \tag{2.71}$$

The normalized load current relative to the rectified source voltage is plotted in Figure 2.14

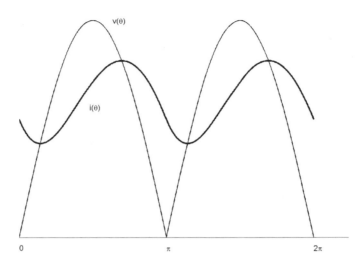

FIGURE 2.14   Load voltage and current of the FW rectifier with R-L load.

Fortunately, there is a direct way to obtain the DC component of the load current without the need to integrate Equation 2.71. The voltage across the R-L load is the FW rectified sine-wave, the average value of which is expressed by Equation 2.32. Since the inductor is a short circuit to the DC component, the average voltage appears directly across the resistance; the average current, therefore, is expressed by Equation 2.33, the diode dissipation by Equation 2.6, and the efficiency by Equation 2.39.

Power Computations

As demonstrated in Example 2.6, the power computations are readily performed by numerical integration in MATLAB. The next example illustrates the power computations for the FW rectifier with R-L load.

EXAMPLE 2.7

Use MATLAB to perform the power computations for the FW rectifier with R-L load with circuit parameters $V_s = 6.3$ VAC, $f = 60$ Hz, $R = 2\ \Omega$, $L = 50$ mH, and $V_f = 0.7$ V. Can the power factor of the circuit be improved with shunt capacitance?

Solution

ON THE CD

The solution is provided by the code in Listing 2.5. The program is available as Example2_7.m in Chapter 2\Examples and Listings on the CD-ROM.

LISTING 2.5 Solution to Example 2.7

```
close all, clear all, clc
Vs = 6.3;
Vm = Vs*sqrt(2);
Vf = 0.7;
y = Vf/Vm;
f = 60;
T = pi;
w = 2*pi*f;
R = 2;
L = 0.05;
q = w*L/R;
phi = atan(q);
th = linspace(0, pi, 1000);
Ith = Vm/R*(sin(2*phi)/(1 - exp(-pi/q))*exp(-th/q) + ...
    cos(phi)*sin(th - phi));
Is = sqrt(1/T*trapz(th, Ith.^2))
a1 = 2/T*trapz(th, Ith.*cos(th));
b1 = 2/T*trapz(th, Ith.*sin(th));
Is1 = sqrt((a1^2 + b1^2)/2);
S = Vs*Is
P = Vm*b1/2
Pdio = Vm*Vf/(pi*R)
Q = -Vm*a1/2
D = Vs*sqrt(Is^2 - Is1^2)
pf = P/S
ph1 = atan(a1/b1);
DPF = cos(ph1)
DF = Is1/Is
eta = 1/(1 + 8/pi*y/(1 - 2*y))
```

Conclusion

The results of the computations performed by the code in Listing 2.5 are as follows:

$$I_{rms} = 2.84 \text{ A}$$
$$S = 17.9 \text{ VA}$$
$$P = 16.1 \text{ W}$$

$P_{DIO} = 0.99$ W

$Q = 0.40$ VAr

$D = 7.76$ VAd

$pf = 0.90$

$DPF = 1.00$

$DF = 0.90$

$\eta = 80.8\%$.

The positive VAr indicates that power factor improvement is possible; however, Q is practically zero and the displacement factor is unity to 3 significant digits. The PFC capacitor is virtually ineffective for this circuit example.

EXERCISE 2.7

Compute the shunt capacitance for Example 2.7. What is the percent improvement in the power factor with this capacitor?

Answers

$C = 27$ μF. The power factor improvement is only 0.025%.

HW Battery Charger with Series Resistance

The simplest battery charging circuit is the HW rectifier. The resistance modeled in the circuit includes the internal resistance of the battery and the secondary winding resistance of the supply transformer. The HW battery charger with series resistance is shown in Figure 2.15.

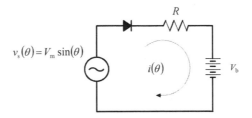

FIGURE 2.15   HW battery charger with series resistance.

Application of Kirchhoff's voltage law around the circuit loop of Figure 2.15 results in

$$V_m \sin\theta = Ri(\theta) + V_b. \qquad (2.72)$$

The solution of Equation 2.72 for the instantaneous current is

$$i(\theta) = \frac{V_m}{R}\sin\theta - \frac{V_b}{R}. \qquad (2.73)$$

The normalized load voltage and load current are illustrated in Figure 2.16.

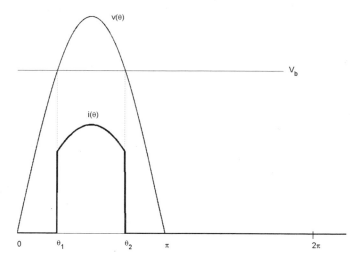

FIGURE 2.16   Battery current of the HW charger with series resistance.

As shown in Figure 2.16, the diode does not conduct current until the source voltage exceeds the battery voltage. With Equation 2.73 set to zero, the solution for the angle at which the diode becomes forward-biased is

$$\theta_1 = \arcsin\left(\frac{V_b}{V_m}\right). \qquad (2.74)$$

By symmetry, the angle at which the diode becomes reverse-biased is

$$\theta_2 = \pi - \theta_1. \qquad (2.75)$$

Average Battery Current

Application of the average value integral to Equation 2.73 yields the average battery current:

$$I_b = \frac{1}{2\pi R} \int_{\theta_1}^{\pi-\theta_1} \left[V_m \sin\theta - V_b \right] d\theta.$$

Evaluation of the average value integral yields

$$I_b = \frac{1}{\pi R} \left[V_m \cos\theta_1 - V_b \left(\frac{\pi}{2} - \theta_1 \right) \right]. \tag{2.76}$$

The power absorbed by the battery is the product of the battery voltage and the average current:

$$P_b = \frac{V_b}{\pi R} \left[V_m \cos\theta_1 - V_b \left(\frac{\pi}{2} - \theta_1 \right) \right]. \tag{2.77}$$

The battery current also flows through the diode; the diode power dissipation, therefore, is

$$P_{DIO} = \frac{V_f}{\pi R} \left[V_m \cos\theta_1 - V_b \left(\frac{\pi}{2} - \theta_1 \right) \right]. \tag{2.78}$$

In the circuits previously considered, the resistance in the circuit was contained entirely in the load. In the HW charger, the series resistance models the losses in the circuit and thus affects the circuit efficiency. Equation 2.10 is readily modified to include the ohmic losses of the HW charger:

$$\eta = \frac{P_b}{P_b + P_R + P_{DIO}}. \tag{2.79}$$

Power Computations

The power computations for the HW charger are illustrated in the next example.

EXAMPLE 2.8

Use MATLAB to perform the power computations for the HW battery charger with circuit parameters $V_s = 75$ VAC, $f = 60$ Hz, $R = 0.7 \ \Omega$, $V_b = 90$ VDC, and $V_f = 1.2$ V.

Solution

ON THE CD The solution is provided by the code in Listing 2.6. The program is available as Example2_6.m in Chapter 2\Examples and Listings on the CD-ROM.

LISTING 2.6 Solution to Example 2.8

```
close all, clear all, clc
Vs = 75;
Vm = Vs*sqrt(2);
Vb = 90;
Vf = 1.2;
f = 60;
T = 2*pi;
w = 2*pi*f;
R = 0.7;
th1 = asin(Vb/Vm);
th2 = pi - th1;
th = linspace(th1, th2, 1024);
Ith = Vm/R*sin(th) - Vb/R;
Is = sqrt(1/T*trapz(th, Ith.^2))
Ib = 1/(pi*R)*(Vm*cos(th1) - Vb*(pi/2 - th1))
a1 = 2/T*trapz(th, Ith.*cos(th));
b1 = 2/T*trapz(th, Ith.*sin(th));
Is1 = sqrt((a1^2 + b1^2)/2);
S = Vs*Is
P = Vm*b1/2
Pdio = Vf*Ib
Q = -Vm*a1/2
D = Vs*sqrt(Is^2 - Is1^2)
pf = P/S
ph1 = atan(a1/b1);
DPF = cos(ph1)
DF = Is1/Is
Pr = Is^2*R
Pb = Vb*Ib
eta = Pb/(Pb + Pr + Pdio)
```

Conclusion

The results produced by the code in Listing 2.6 are as follows:

$I_{rms} = 7.04$ A

$I_b = 2.70$ A

$S = 528 \text{ VA}$

$P = 278 \text{ W}$

$P_{\text{DIO}} = 3.24 \text{ W}$

$Q = 0$

$D = 449 \text{ VAd}$

$pf = 0.53$

$DPF = 1$

$DF = 0.53$

$\eta = 86.5\%.$

The average power computed is the sum of the power dissipated in the series resistance (34.6 W) and the power absorbed by the battery (243 W) and represents the power delivered by the source. The power factor cannot be improved with shunt capacitance because the reactive power is zero. The poor power factor is due entirely to current harmonics above the fundamental.

Exercise 2.8

Derive the efficiency expression for the HW charger in terms of the series resistance and the battery and diode voltages.

Answer

$$\eta = \frac{V_b}{V_b + I_{\text{RMS}}^2 R / I_b + V_f}.$$

Full-Wave Battery Charger with Series Resistance

The FW battery charger with series resistance is illustrated in Figure 2.17.

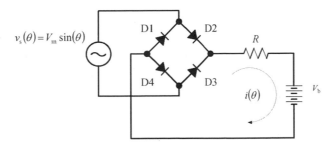

FIGURE 2.17   FW battery charger with series resistance.

The instantaneous battery current of the FW charger is expressed by the same equation as the HW charger; the difference, however, is that the current of the FW circuit has period π rather than 2π. The battery current is plotted in Figure 2.18 along with the FW rectified source voltage.

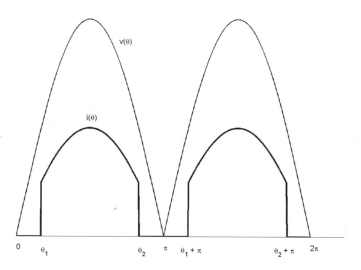

FIGURE 2.18   Instantaneous battery current of the FW charger with series resistance.

With a period of π, the average battery current of the FW charger is twice that of the HW charger:

$$I_b = \frac{2}{\pi R}\left[V_m \cos\theta_1 - V_b\left(\frac{\pi}{2} - \theta_1\right)\right].\tag{2.80}$$

The power absorbed by the battery is thus twice that of the HW charger:

$$P_b = \frac{2V_b}{\pi R}\left[V_m \cos\theta_1 - V_b\left(\frac{\pi}{2} - \theta_1\right)\right].\tag{2.81}$$

The diode dissipations of the FW charger are expressed by Equation 2.78 because the period of the diode currents is 2π. Equation 2.79 is readily modified to include the losses of the additional diodes of the FW circuit:

$$\eta = \frac{P_o}{P_o + P_R + 4P_{DIO}}.\tag{2.82}$$

Power Computations

The next example illustrates the power computations for the FW charger with series resistance.

EXAMPLE 2.9

Write a MATLAB program to compute the power and related functions for a full-wave charger with circuit parameters $V_s = 75$ VAC, $f = 60$ Hz, $R = 0.7\ \Omega$, $V_b = 90$ VDC, and $V_f = 1.2$ V.

Solution

The solution is provided by the code in Listing 2.6 with the period T changed to π and the average battery current and efficiency computations modified in accordance with Equations 2.80 and 2.82, respectively.

Conclusion

The results produced by the modified code are as follows:

$I_{rms} = 9.95$ A

$I_b = 5.40$ A

$S = 746$ VA

$P = 556$ W

$P_{DIO} = 3.24$ W

$Q = 0$

$D = 498$ VAd

$pf = 0.74$

$DPF = 1$

$DF = 0.74$

$\eta = 85.5\%$.

A comparison of the computations in Example 2.9 with those of Example 2.8 shows that the source power of the FW charger is indeed twice that of the HW charger. The series resistance dissipation is also doubled (69.3 W), as is the battery power (486 W).

Although the circuit power factor cannot be improved with shunt capacitance ($Q = 0$), the power factor of the FW circuit is significantly better than that of the HW circuit.

EXERCISE 2.9

Determine the RMS source voltage required to charge a 144 VDC battery pack with a series resistance of 0.3 Ω at a rate of 20 A. If this voltage is supplied by a transformer, what is the kilovoltampere (kVA) rating?

Answers

$V_s = 120$ and $S = 4.5$ kVA.

HW Battery Charger with Series R-L Impedance

As demonstrated earlier, a physical power circuit has sufficient inductance to affect the source current waveform. The HW charger with series R-L impedance is illustrated in Figure 2.19.

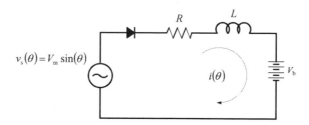

FIGURE 2.19 HW charger with series R-L impedance.

The summation of voltages around the output loop of the circuit in Figure 2.19 yields the first-order differential equation

$$V_m \sin\theta = iR + \omega L \frac{di}{d\theta} + V_b. \tag{2.83}$$

In terms of the inductor quality factor, Equation 2.83 is expressed as

$$\frac{di}{d\theta} + \frac{i}{q} = \frac{V_m}{qR}\sin\theta - \frac{V_b}{qR}. \tag{2.84}$$

By the principle of superposition, the complete response of Equation 2.84 is Equation 2.56 minus the battery current:

$$i(\theta) = A_0 e^{-\theta/q} + \frac{V_m}{R}\cos\phi\sin(\theta-\phi) - \frac{V_b}{R}. \tag{2.85}$$

As with the other battery chargers, current does not conduct through the circuit until the source voltage exceeds the battery voltage. The initial condition, therefore, is

$$i(\theta_1) = 0,\qquad(2.86)$$

in which θ_1 is expressed by Equation 2.74. Application of the initial condition to Equation 2.85 yields the expression

$$A_0 e^{-\theta_1/q} + \frac{V_m}{R}\cos\phi\sin(\theta_1 - \phi) - \frac{V_b}{R} = 0,$$

for which the solution for A_0 is

$$A_0 = \left[\frac{V_b}{R} - \frac{V_m}{R}\cos\phi\sin(\theta_1 - \phi)\right]e^{\theta_1/q}.\qquad(2.87)$$

Let the ratio of the battery voltage to the peak source voltage be defined as

$$\gamma = \frac{V_b}{V_m}.\qquad(2.88)$$

Substitution of Equations 2.87 and 2.88 into Equation 2.85 produces the expression for the instantaneous current of the HW charger with series R-L impedance:

$$i(\theta) = \frac{V_m}{R}\left\{\left[\gamma - \cos\phi\sin(\theta_1 - \phi)\right]e^{-(\theta-\theta_1)/q} + \cos\phi\sin(\theta - \phi) - \gamma\right\}.\qquad(2.89)$$

The normalized instantaneous battery current is plotted in Figure 2.20 along with the HW rectified source voltage waveform.

The instantaneous battery current is zero at the extinction angle θ_2. Substitution of $\theta = \theta_2$ with Equation 2.89 set to zero, the extinction angle and cut-in angle are related by

$$\left[\gamma - \cos\phi\sin(\theta_1 - \phi)\right]e^{\theta_1/q} = \left[\gamma - \cos\phi\sin(\theta_2 - \phi)\right]e^{\theta_2/q}.\qquad(2.90)$$

The next example illustrates a numerical solution technique for Equation 2.90.

EXAMPLE 2.10

Modify the function in Listing 2.3 to compute the values of θ_1 and θ_2 from q and γ. Use the function to compute the values of θ_1 and θ_2 for a half-wave charger with

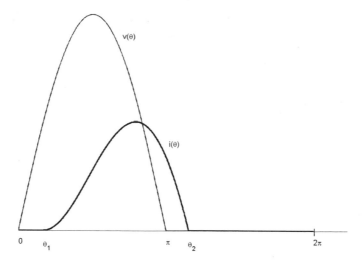

FIGURE 2.20   Battery current of the HW charger with series R-L impedance.

circuit parameters $V_s = 120$ VAC, $f = 60$ Hz, $R = 1.5\ \Omega$, $L = 20$ mH, and $V_b = 90$ VDC.

Solution

ON THE CD

The solution is provided by the MATLAB code in Listings 2.7 and 2.8. The function in Listing 2.7 is available as th12solv.m in Chapter 2\Toolbox on the CD-ROM. The code in Listing 2.8 is contained in file Example2_10.m in Chapter 2\Examples and Listings.

LISTING 2.7   Function to Compute θ_1 and θ_2 from q and γ

```
function [th1, th2] = th12solv(q, gma)
phi = atan(q);
LHS = 1;
th1 = asin(gma);
th2 = pi;
inc = pi/2;
while(abs(LHS) > 1E-6),
    LHS = (gma - cos(phi)*sin(th1 - phi))*exp(th1/q) - ...
          (gma - cos(phi)*sin(th2 - phi))*exp(th2/q);
    if LHS > 0,
        th2 = th2 + inc;
```

```
        else
            th2 = th2 - inc;
            inc = inc/2;
            th2 = th2 + inc;
        end
    end
end
```

LISTING 2.8   Solution to Example 2.10

```
close all, clear all, clc
Vs = 120;
Vm = Vs*sqrt(2);
f = 60;
w = 2*pi*f;
R = 1.5;
L = 0.02;
Vb = 90;
q = w*L/R;
phi = atan(q);
gma = Vb/Vm;
[th1, th2] = th12solv(q, gma)
```

Conclusion

The function is readily modified to add additional input and output arguments and to include the coded forms of Equations 2.74 and 2.90. The eighth line of the listing makes use of the ellipses (…) to continue the lengthy expression onto the next line. The output of the program in Listing 2.8 is $\theta_1 = 0.56$ radians and $\theta_2 = 3.56$ radians.

EXERCISE 2.10

Use *trapz* and Equation 2.89 to compute the average battery current of the circuit in Example 2.10.

Answer

$I_b = 3.14$ A.

Average Battery Current

Application of the average value integral to Equation 2.89 results in

$$I_b = \frac{1}{2\pi} \int_{\theta_1}^{\theta_2} \left\{ \left[\frac{V_b}{R} - \frac{V_m}{R} \cos\phi \sin(\theta_1 - \phi) \right] e^{-(\theta - \theta_1)/q} + \frac{V_m}{R} \cos\phi \sin(\theta - \phi) - \frac{V_b}{R} \right\} d\theta. \quad (2.91)$$

The solutions to the integrals of Equation 2.91 are as follows:

$$\int_{\theta_1}^{\theta_2} e^{-(\theta-\theta_1)/q} \, d\theta = q\left[1 - e^{-(\theta_2-\theta_1)/q} \right]. \tag{2.92}$$

$$\int_{\theta_1}^{\theta_2} \sin(\theta-\phi) \, d\theta = \cos\left(\theta_1 - \phi\right) - \cos\left(\theta_2 - \phi\right). \tag{2.93}$$

$$\int_{\theta_1}^{\theta_2} d\theta = \theta_2 - \theta_1. \tag{2.94}$$

The difference between the angles at which the diode begins and ceases to conduct is called the *conduction angle*:

$$\theta_c = \theta_2 - \theta_1. \tag{2.95}$$

With Equations 2.92 to 2.95, the average battery current is expressed as

$$I_b = \frac{V_m}{2\pi R}\left\{ q\left[\gamma - \cos\phi\sin(\theta_1 - \phi)\right]\left[1 - e^{-\theta_c/q} \right] + \cos\phi\left[\cos(\theta_1 - \phi) - \cos(\theta_2 - \phi)\right] - \gamma\theta_c \right\}. \tag{2.96}$$

Equation (2.96) is admittedly complicated; however, MATLAB makes the computation tractable.

EXAMPLE 2.11

Determine the RMS source voltage required to supply 10 A to the battery in Example 2.10.

Solution

While it is possible to write a program to solve Equation 2.96 iteratively for V_m, a different approach is presented in this example. Often a solution is obtained much faster with a program that simply performs a calculation with an estimate of the unknown value. The estimate is refined and the program is executed repeatedly until the desired results are obtained. The solution code for this example appears in Listing 2.9 and is available on the CD-ROM as Example2_11.m in Chapter 2\Examples and Listings.

ON THE CD

LISTING 2.9   Program to Compute Average Battery Current from Equation 2.96

```
close all, clear all, clc
Vs = 190;
Vm = Vs*sqrt(2);
f = 60;
w = 2*pi*f;
R = 1.5;
L = 0.02;
Vb = 90;
q = w*L/R;
phi = atan(q);
gma = Vb/Vm;
[th1, th2] = th12solv(q, gma);
thc = th2 - th1;
Ib = Vm/(2*pi*R)*(q*(gma - cos(phi)*sin(th1 - phi))*...
(1 - exp(-thc/q)) + cos(phi)*(cos(th1 - phi) - ...
cos(th2 - phi)) - gma*thc)
```

Conclusion

With the program in Listing 2.9, the required RMS source voltage that produces a 10 A battery current is readily found to be 190 VRMS.

EXERCISE 2.11

Modify the code in Listing 2.9 to solve for V_m iteratively given the value of battery current I_b.

Diode Dissipation and Efficiency

Since the diode and battery current are the same, the diode dissipation is simply the product of the forward voltage and the average battery current expressed by Equation 2.96. Furthermore, the efficiency of the HW charger with series R-L impedance is computed with Equation 2.79 since no power is absorbed by the inductance.

Power Computations

The next example illustrates the power computations for the HW charger with series R-L impedance.

EXAMPLE 2.12

Write a MATLAB program to perform the power computations for the HW charger with series R-L impedance with the circuit parameters of Example 2.10. Is it possible to improve the power factor with shunt capacitance?

Solution

ON THE CD

The solution is provided by the MATLAB code in Listing 2.10 and is available on the CD-ROM as Example2_12.m in Chapter 2\Examples and Listings.

LISTING 2.10   Power Computations for the HW Charger with Series R-L Impedance

```
close all, clear all, clc
T = 2*pi;
Vs = 120;
Vm = Vs*sqrt(2);
f = 60;
w = 2*pi*f;
R = 1.5;
L = 0.02;
Vb = 90;
q = w*L/R;
phi = atan(q);
v = Vb/Vm;
[th1, th2] = th12solv(q, v);
th = linspace(th1, th2, 1024);
Ith = Vm/R*((v - cos(phi)*sin(th1 - phi))*exp(-(th - th1)/q) + ...
        cos(phi)*sin(th - phi) - v);
Ib = 1/T*trapz(th, Ith);
Is = sqrt(1/T*trapz(th, Ith.^2))
a1 = 2/T*trapz(th, Ith.*cos(th));
b1 = 2/T*trapz(th, Ith.*sin(th));
S = Vs*Is
P = Vm*b1/2
Q = -Vm*a1/2
Is1 = sqrt((a1^2 + b1^2)/2);
D = Vs*sqrt(Is^2 - Is1^2)
ph1 = atan(a1/b1);
pf = P/S
DPF = cos(ph1)
DF = Is1/Is
```

Conclusion

The results produced by the code in Listing 2.10 are as follows:

$I_s = 5.31$ A

$S = 637$ VA

$P = 325$ W

$Q = 303$ VAr

$D = 457$ VAd

$pf = 0.51$

$DPF = 0.73$

$DF = 0.70.$

The positive value of Q indicates that shunt capacitance will improve the power factor; however, the distortive volt-amperes are greater than both the real and reactive powers. The poor power factor is largely due to current harmonics.

EXERCISE 2.12

Compute the shunt capacitance required to neutralize the reactive power in Example 2.12. Compute the improved power factor.

Answers

$C = 55.7 \ \mu F$ and $pf_C = 0.58.$

FW Battery Charger with Series R-L Impedance

The FW charger with series R-L impedance is the most common battery charger configuration. The circuit is more costly but offers greater charging currents and better power factors. The circuit is illustrated in Figure 2.21.

The differential equation and general solution for the FW charger current are the same as for the HW charger; however, there are two modes of operation for this circuit. If the inductance is of sufficient magnitude to sustain the current flow throughout the first half-cycle of the source, then the battery current is continuous. If the inductance is not large enough, or if the battery voltage exceeds a certain threshold, the current becomes discontinuous.

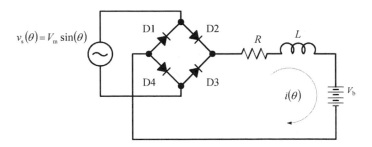

FIGURE 2.21 FW charger with series R-L impedance.

Continuous Conduction Mode (CCM)

When the battery current is continuous, the voltage across the combined R-L-V_b load is the FW rectified sine-wave with an average value computed with Equation 2.32. The average resistor voltage is the difference between this average value and the battery voltage:

$$V_R = \frac{2V_m}{\pi} - V_b. \tag{2.97}$$

Therefore, the average current that charges the battery is

$$I_b = \frac{2V_m}{\pi R} - \frac{V_b}{R}. \tag{2.98}$$

Instantaneous CCM Current

In CCM, the boundary condition $i(0) = i(\pi)$ must be satisfied. Substitution of $\theta = 0$ into Equation 2.85 yields the expression

$$i(0) = A_0 - \frac{V_m}{R}\cos\phi\sin\phi - \frac{V_b}{R}. \tag{2.99}$$

Similarly, substitution of $\theta = \pi$ yields

$$i(\pi) = A_0 e^{-\pi/q} + \frac{V_m}{R}\cos\phi\sin\phi - \frac{V_b}{R}. \tag{2.100}$$

Application of the boundary condition to Equations 2.99 and 2.100 produces the expression

$$A_0 - \frac{V_m}{R}\cos\phi\sin\phi = A_0 e^{-\pi/q} + \frac{V_m}{R}\cos\phi\sin\phi. \tag{2.101}$$

The solution of Equation 2.101 for A_0 is

$$A_0 = \frac{V_m}{R}\frac{\sin 2\phi}{1 - e^{-\pi/q}}. \tag{2.102}$$

Substitution of Equation 2.102 into Equation 2.85 yields the instantaneous continuous current of the FW charger with R-L impedance:

$$i(\theta) = \frac{V_m}{R}\left[\frac{\sin 2\phi}{1 - e^{-\pi/q}}e^{-\theta/q} + \cos\phi\sin(\theta - \phi) - \gamma\right]. \tag{2.103}$$

The normalized continuous current is plotted in Figure 2.22 along with the FW rectified source voltage.

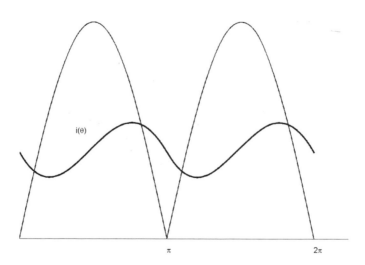

FIGURE 2.22   Continuous current of the FW charger with R-L impedance.

Diode Dissipation and Efficiency

The diodes in the FW charger have the same duty ratio as the HW charger. The average diode current, therefore, is half the average battery current expressed by Equation 2.101. Consequently, the diode dissipation is

$$P_{\text{DIO}} = \frac{V_f V_m}{\pi R} - \frac{V_f V_b}{2R}. \qquad (2.104)$$

The losses of the FW charger occur in the four diodes and the series resistance. The efficiency, therefore, is expressed as

$$\eta = \frac{P_b}{P_b + P_R + 4P_{\text{DIO}}}. \qquad (2.105)$$

Power Computations

The power computations for the FW charger with series R-L impedance are illustrated in the next example.

EXAMPLE 2.13

Compute the source power and related power functions for a full-wave charger with R-L impedance with continuous battery current and with circuit parameters as in Example 2.12.

Solution

ON THE CD

The solution is provided by the code in Listing 2.11 and is available as Example2_13.m in Chapter 2\Examples and Listings on the CD-ROM.

LISTING 2.11   Power Computations for a CCM FW Charger with Series R-L Impedance

```
close all, clear all, clc
T = pi;
Vs = 120;
Vm = Vs*sqrt(2);
Vb = 90;
f = 60;
w = 2*pi*f;
L = 0.02;
R = 1.5;
q = w*L/R;
phi = atan(q);
v = Vb/Vm;
th = linspace(0, pi, 1024);
```

```
Ith = Vm/R*(sin(2*phi)/(1 - exp(-pi/q))*exp(-th/q) + ...
         cos(phi)*sin(th - phi) - v);
Is = sqrt(1/T*trapz(th, Ith.^2))
S = Vs*Is
a1 = 2/T*trapz(th, Ith.*cos(th));
b1 = 2/T*trapz(th, Ith.*sin(th));
P = Vm*b1/2
Q = -Vm*a1/2
Is1 = sqrt((a1^2 + b1^2)/2);
D = Vs*sqrt(Is^2 - Is1^2)
pf = P/S
ph1 = atan(a1/b1);
DPF = cos(ph1)
DF = Is1/Is
```

Conclusion

The results produced by the code in Listing 2.10 are as follows:

$I_s = 12.5$ A

$S = 1.50$ kVA

$P = 1.32$ kW

$Q = 358$ VAr

$D = 621$ VAd

$pf = 0.88$

$DPF = 0.96$

$DF = 0.91.$

The results show that the power factor of the FW circuit is much better than that of the HW circuit. Power factor improvement is possible with shunt capacitance; however, as with the HW charger, the source current has appreciable harmonic content.

EXERCISE 2.13

Compute the power absorbed by the battery and the power dissipated in the series resistance of the circuit in Example 2.13. Compute the diode dissipation for a forward voltage of 1.1 V, the shunt capacitance required to neutralize the reactive power, and the improved power factor.

Answers

$P_b = 1.08$ kW
$P_R = 234$ W
$P_{DIO} = 6.6$ W
$\eta = 81\%$
$C = 66$ μF
$pf = 0.90.$

Discontinuous Conduction Mode (DCM)

If the inductance is insufficient to sustain the battery current up to the next half-cycle of the source voltage, the FW rectifier becomes reverse-biased and the battery current becomes discontinuous. The initial condition in this case is the same as that of the HW charger, and the instantaneous battery current is expressed by Equation 2.89 with a period of π. The average battery current of the FW circuit is thus twice that expressed by Equation 2.96:

$$I_b = \frac{V_m}{\pi R}\left\{ q\left[\gamma - \cos\phi \sin(\theta_1 - \phi)\right]\left[1 - e^{-\theta_c/q} \right] + \cos\phi\left[\cos(\theta_1 - \phi) - \cos(\theta_2 - \phi)\right] - \gamma\theta_c \right\}. \quad (2.106)$$

When the current is discontinuous, the voltage across the R-L-V_b load is no longer the FW rectified sine waveform. During the conduction interval, the output voltage is the sum of the battery voltage and the voltages across the resistor and the inductor. When the diode bridge is reverse-biased, the battery current is zero and the output voltage is equal to the battery voltage. The output voltage of the rectifier in DCM is thus

$$v(\theta) = i(\theta)R + \omega L\frac{di(\theta)}{d\theta} + V_b, \quad \theta_1 \leq \theta \leq \theta_2 \quad (2.107)$$

and

$$v(\theta) = V_b, \quad 0 \leq \theta \leq \theta_1 \text{ and } \theta_2 \leq \theta \leq \theta_1 + \pi. \quad (2.108)$$

Substitution of Equation 2.89 and its derivative into Equation 2.107 yields the output voltage expression during the conduction interval:

$$v(\theta) = V_m \cos\phi \sin(\theta - \phi) + V_m \sin\phi \cos(\theta - \phi), \quad \theta_1 \leq \theta \leq \theta_2. \quad (2.109)$$

The normalized rectifier output voltage waveform and the discontinuous battery current are plotted in Figure 2.23.

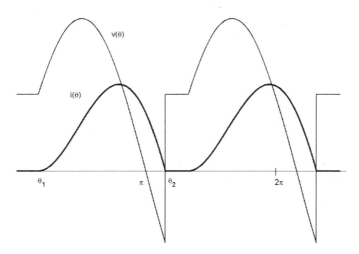

FIGURE 2.23   Rectifier output voltage with discontinuous battery current.

The CCM and DCM circuit operations are both advantageous modes for charging batteries. In CCM the average battery current is larger than in DCM and thus more charge is delivered to the battery. As the battery voltage increases as charge accumulates in the battery, the charger moves into DCM operation when the battery voltage exceeds the average value of the FW-rectified sine-wave. In DCM, because the rectifier bridge is periodically reverse-biased, the conduction angle is shorter and results in reduced average battery current. As the battery voltage continues to increase, the conduction angle further decreases and reduces the average battery current even more. The move from CCM to DCM thus performs charge regulation.

EXAMPLE 2.14

A certain 12 V battery has a terminal voltage of 10 V when fully discharged and a full-charge voltage of 13.5 V. A battery pack is constructed of 12 of these batteries in series. If the rectifier supply voltage is 120 VAC, compute the average battery current at both the fully discharged and fully charged states. The series resistance is 0.3 Ω, and the series inductance is 0.25 mH.

Solution

The average value of the rectified sine-wave is approximately 108 V. The terminal voltage of the discharged battery pack is $(12)(10) = 120$ VDC; the voltage of the fully charged pack is $(12)(13.5) = 162$ VDC. Since the pack voltage is always greater than the average value of the rectified sine-wave, the charger always operates in

DCM. The code that performs the average battery current computation is shown in Listing 2.12 and is available on the CD-ROM as Example2_14.m in folder Chapter 2\Examples and Listings.

LISTING 2.12 Average Battery Current Computation for Example 2.14

```
close all, clear all, clc
T = pi;
Vs = 120;
Vm = Vs*sqrt(2);
Vb = 120;
f = 60;
w = 2*pi*f;
L = 0.25E-3;
R = 0.3;
q = w*L/R;
phi = atan(q);
v = Vb/Vm;
[th1, th2] = th12solv(q, v);
thc = th2 - th1;
Ib = Vm/(pi*R)*(q*(v - cos(phi)*sin(th1 - phi))* ...
(1 - exp(-thc/q)) + cos(phi)*(cos(th1 - phi)-cos(th2 - phi)) -
 v*thc)
```

Conclusion

The code in Listing 2.11 computes a value of 50 A for the average battery current when the battery voltage is 120 VDC. At 162 VDC, the average battery current is 2.4 A. The inherent current tapering is a very cost-effective method of charge control.

It is generally desirable to charge a depleted battery at a high rate to hasten the charging process and then gradually taper the current to near zero toward the end of the charging cycle.

EXERCISE 2.14

Compute the average battery current of the discharged battery pack if the series inductance is $L = 1$ mH.

Answer

$I_b = 32$ A.

Power Computations

The next example illustrates the power computations for the DCM charger.

EXAMPLE 2.15

Write a MATLAB program to perform the power computations for the FW DCM charger with the circuit parameters from Example 2.10. Is it possible to improve the power factor with shunt capacitance?

Solution

ON THE CD

The solution code is given by Listing 2.9, with the period T changed to π. The modified code is available as Example2_15.m in Chapter 2\Examples and Listings on the CD-ROM. The computed results are as follows:

$$I_s = 7.51 \text{ A}$$
$$S = 901 \text{ VA}$$
$$P = 649 \text{ W}$$
$$Q = 605 \text{ VAr}$$
$$D = 155 \text{ VAd}$$
$$pf = 0.72$$
$$DPF = 0.73$$
$$DF = 0.99.$$

Conclusion

The division of the period by 2 results in a doubling of the real and reactive powers as computed in Example 2.12. The increased conduction time provided by FW rectification reduces the VAd content and improves the power factor. The large positive VAr component makes the circuit an excellent candidate for power factor improvement with shunt capacitance.

EXERCISE 2.15

Compute the power absorbed by the battery and the power dissipated in the series resistance of the circuit in Example 2.15. Compute the diode dissipation and the circuit efficiency for a forward voltage of 0.8 V. Compute the shunt capacitance to neutralize the reactive power along with the improved power factor.

Answers

$$P_b = 565 \text{ W}$$
$$P_R = 84.5 \text{ W}$$
$$P_{DIO} = 2.5 \text{ W}$$
$$\eta = 86\%$$

$C = 111 \, \mu\text{F}$

$pf_C = 0.97.$

RECTIFIER CIRCUITS WITH LOAD CAPACITANCE

The rectifier circuits considered thus far are largely used in applications where the average current is of interest; the average voltage is of secondary interest. The battery charger is the best example because the average current contributes charge to the battery.

When an average voltage is of primary importance, a filter capacitance is placed in parallel with the resistive load to smooth the output voltage variations and increase the average output voltage. The resistive-capacitive (R-C) load behaves somewhat like the battery in the charger circuits in that the rectifier is periodically reverse-biased.

HW Rectifier with R-C Load

Figure 2.24 shows the HW rectifier with resistive load with a capacitor filter.

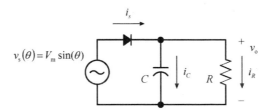

FIGURE 2.24 HW rectifier with R-C load.

Source Current

When the source voltage exceeds the capacitor voltage, the diode is forward-biased and the source current flows into the capacitor and the resistor. When the source voltage falls below the capacitor voltage, the diode becomes reverse-biased and the source is disconnected from the load. The capacitor then discharges into the load resistance until the capacitor voltage is exceeded once again by the source voltage. The HW rectified source, the source current, and the output voltage are plotted in Figure 2.25.

As shown in Figure 2.25, the output voltage follows the source voltage during the interval $\theta_1 \leq \theta \leq \theta_2$, while the diode is forward-biased. When the source current is interrupted, the output voltage exponentially decays until the start of the next source cycle.

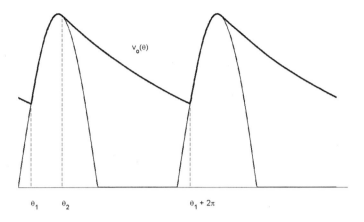

FIGURE 2.25   Output voltage of the HW rectifier with R-C load.

With reference to Figure 2.24, the source current is expressed as the sum of the capacitor and resistor currents:

$$i_s = i_C + i_R.$$ (2.110)

Substitution of the respective current expressions into Equation (2.110) yields the differential equation

$$i_s(\theta) = \omega C \frac{dv_o(\theta)}{d\theta} + \frac{v_o(\theta)}{R}.$$ (2.111)

While the diode is forward-biased, the output voltage is

$$v_o(\theta) = V_m \sin\theta, \quad \theta_1 \le \theta \le \theta_2.$$ (2.112)

Substitution of Equation 2.112 into Equation 2.111 results in the expression for the source current while the diode is forward-biased:

$$i_s(\theta) = \omega C V_m \cos\theta + \frac{V_m}{R} \sin\theta, \quad \theta_1 \le \theta \le \theta_2.$$ (2.113)

The quality factor of the capacitor is defined as

$$q = \omega RC.$$ (2.114)

In terms of the capacitor quality factor, the instantaneous source current is

$$i_s(\theta) = \frac{V_m}{R}\left[q\cos\theta + \sin\theta \right], \quad \theta_1 \le \theta \le \theta_2. \tag{2.115}$$

The source current is zero at $\theta = \theta_2$ when the diode is reverse-biased:

$$\frac{V_m}{R}\left[q\cos\theta_2 + \sin\theta_2 \right] = 0. \tag{2.116}$$

As indicated in Figure 2.25, angle θ_2 lies in the second quadrant. Therefore, the solution to Equation 2.116 for θ_2 is

$$\theta_2 = \pi - \arctan(q). \tag{2.117}$$

Output Voltage

As expressed by Equation 2.112, the output voltage follows the sinusoidal source voltage while the diode is forward-biased. When the diode is reverse-biased, the source current is interrupted and Equation (2.111) becomes

$$\frac{dv_o(\theta)}{d\theta} + \frac{v_o(\theta)}{q} = 0. \tag{2.118}$$

The general solution to Equation 2.118 is

$$v_o(\theta) = Ae^{-\theta/q}. \tag{2.119}$$

The initial condition for Equation 2.119 is provided by Equation 2.112 with $\theta = \theta_2$:

$$V_m \sin\theta_2 = Ae^{-\theta_2/q}. \tag{2.120}$$

The solution of Equation 2.120 for A is

$$A = V_m e^{\theta_2/q} \sin\theta_2. \tag{2.121}$$

Substitution of Equation 2.121 into Equation 1.119 yields the expression for the output voltage while the diode is reverse-biased:

$$v_o(\theta) = V_m \sin(\theta_2) e^{-(\theta-\theta_2)/q}, \quad \theta_2 \le \theta \le \theta_1 + 2\pi. \tag{2.122}$$

Cut-in and Cut-out Angles

As indicated in Figure 2.25, the output voltage expressed by Equation 2.122 is equal to that expressed by Equation 2.112 when $\theta = \theta_1 + 2\pi$:

$$V_m \sin\left(\theta_2\right)e^{-(\theta_1 + 2\pi - \theta_2)/q} = V_m \sin\left(\theta_1 + 2\pi\right). \tag{2.123}$$

A simplification and rearrangement of Equation 2.123 relates the *cut-in* angle θ_1 to the *cut-out* angle θ_2:

$$e^{(\theta_2 - 2\pi)/q} \sin\theta_2 = e^{\theta_1/q} \sin\theta_1. \tag{2.124}$$

Equation 2.124 is another transcendental equation that must be solved numerically.

EXAMPLE 2.16

Modify the function in Listing 2.3 to solve Equation 2.124 for θ_1.

Solution

ON THE CD

The solution code is provided in Listing 2.13 and is available on the CD-ROM as th12hwrc.m in Chapter 2\Toolbox.

LISTING 2.13   Function to Solve Equation 2.122 for the Cut-in Angle

```
function [th1, th2] = th12hwrc(q)
th2 = pi - atan(q);
LHS = 1;
th1 = 0;
inc = pi/4;
while(abs(LHS) > 1E-6),
    LHS = exp((th2 - 2*pi)/q)*sin(th2) - exp(th1/q)*sin(th1);
    if LHS > 0,
        th1 = th1 + inc;
    else
        th1 = th1 - inc;
        inc = inc/2;
        th1 = th1 + inc;
    end
end
```

Conclusion

The function begins the search for a solution at zero since θ_1 lies in the first quadrant. The function also returns the value of θ_2.

EXERCISE 2.16

A half-wave rectifier with R-C load is required to have a quality factor of 10. Compute the capacitance and the conduction angle with a source frequency of 60 Hz and a load resistance of 72 Ω.

Answers

$C = 368 \ \mu F$ and $\theta_c = 59.6°$.

Average Output Voltage and Ripple Voltage

As stated earlier, the purpose of the filter capacitor is to generate a DC output voltage with limited variation. Application of the average value integral to Equations 2.112 and 2.122 produces

$$V_o = \frac{1}{2\pi}\left[\int_{\theta_1}^{\theta_2} V_m \sin\theta \, d\theta + \int_{\theta_2}^{\theta_1 + 2\pi} V_m \sin(\theta_2) e^{-(\theta - \theta_2)/q} d\theta\right]. \tag{2.125}$$

Evaluation of the integrals in Equation (2.125) yields the average output voltage expression

$$V_o = \frac{V_m}{2\pi}\left\{ \cos\theta_1 - \cos\theta_2 + q\sin\theta_2\left[1 - e^{-(2\pi - \theta_c)/q}\right]\right\}. \tag{2.126}$$

The ripple voltage is defined as the difference between the maximum and minimum values of the instantaneous output voltage. As illustrated in Figure 2.25, the maximum output voltage is the peak value of the source voltage; the minimum occurs at the cut-in angle. Thus, the expression for the peak-to-peak ripple voltage is

$$V_r = V_m\left(1 - \sin\theta_1\right). \tag{2.127}$$

The next example illustrates an application of Equations 2.126 and 2.127.

EXAMPLE 2.17

Compute the average output voltage and the ripple voltage of the circuit in Exercise 2.16 if the peak source voltage is $V_m = 120\sqrt{2}$. Compare these values with those of a half-wave rectifier with a resistive load, that is, the same circuit without the capacitor.

Solution

ON THE CD

The solution code is provided in Listing 2.14 and as file Example2_17.m in folder Chapter 2\Examples and Listings on the CD-ROM.

LISTING 2.14 Solution to Example 2.17

```
close all, clear all, clc
Vm = 120*sqrt(2);
q = 10;
[th1, th2] = th12hwrc(q);
Vr = Vm*(1 - sin(th1))
thc = th2 - th1;
Vo = Vm/(2*pi)*(cos(th1) - cos(th2) + ...
    q*sin(th2)*(1 - exp(-(2*pi - thc)/q)))
```

Conclusion

The average output voltage with the capacitor in the circuit is 134 VDC; the ripple voltage is 69.7 V. Without the capacitor, the average output voltage is only 54 VDC. The capacitor thus greatly increases the DC component of the output voltage. The ripple voltage of the circuit without the capacitor is equal to the peak source voltage (170 V) because the minimum output voltage is zero. The capacitor thus greatly increases the DC component of the output voltage and significantly decreases the output ripple voltage.

EXERCISE 2.17

For the circuit in Example 2.17, determine the value of capacitance required to produce an average output voltage of 150 VDC.

Answer

$C = 774\ \mu F$.

Diode Dissipation and Efficiency

Application of the average value integral to the source current yields the average diode current:

$$I_{\text{DIO}} = \frac{1}{2\pi} \int_{\theta_1}^{\theta_2} \frac{V_m}{R} \Big[q\cos\theta + \sin\theta \Big] d\theta. \tag{2.128}$$

Evaluation of the integrals in Equation 2.128 results in

$$I_{\text{DIO}} = \frac{V_{\text{m}}}{2\pi R}\left\{ q\left[\sin\theta_2 - \sin\theta_1\right] + \cos\theta_1 - \cos\theta_2 \right\}. \tag{2.129}$$

Since the capacitor is modeled as ideal, the diode is the only loss element in the circuit. The efficiency, therefore, is dictated by Equation 2.10 and the diode dissipation is expressed by

$$P_{\text{DIO}} = \frac{V_{\text{f}} V_{\text{m}}}{2\pi R}\left\{ q\left[\sin\theta_2 - \sin\theta_1\right] + \cos\theta_1 - \cos\theta_2 \right\}. \tag{2.130}$$

Capacitor Ripple Current

In addition to the value of capacitance, the current handling capability of the filter capacitor must be specified. The maximum RMS current specified for an electrolytic capacitor is known as the *ripple current* rating. To determine the RMS capacitor current, the instantaneous capacitor current is first formulated as:

$$i_C(\theta) = \omega C \frac{dv_{\text{o}}(\theta)}{d\theta}. \tag{2.131}$$

Substitution of Equations 2.112 and 2.122 into Equation 2.131 yields the expression for the instantaneous capacitor current:

$$i_C(\theta) = \begin{cases} \dfrac{qV_{\text{m}}}{R}\cos\theta, & \theta_1 \leq \theta \leq \theta_2 \\[2mm] -\dfrac{V_{\text{m}}}{R}\sin\left(\theta_2\right)e^{-(\theta-\theta_2)/q}, & \theta_2 \leq \theta \leq \theta_1 + 2\pi \end{cases}. \tag{2.132}$$

The normalized instantaneous capacitor current is plotted in Figure 2.26.

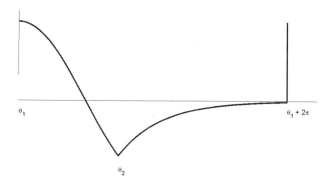

FIGURE 2.26 Instantaneous capacitor current.

The next example illustrates the computation of the capacitor ripple current.

EXAMPLE 2.18

Write a MATLAB program to compute the capacitor ripple current of the circuit in Example 2.17.

Solution

The solution program is provided in Listing 2.15 and as Example2_18.m in Chapter 2\Examples and Listings on the CD-ROM.

LISTING 2.15   Program to Compute RMS Capacitor Current of the HW Rectifier with R-C Load

```
close all, clear all, clc
T = 2*pi;
Vs = 120;
Vm = Vs*sqrt(2);
q = 10;
R = 72;
[th1, th2] = th12hwrc(q);
thc = th2 - th1;
N = 1024;
Na = fix(N*thc/T);
tha = linspace(th1, th2, Na);
Nb = ceil(N*(1 - thc/T));
thb = linspace(th2, th1 + T, Nb);
Ica = q*Vm/R*cos(tha);
Icb = -Vm/R*sin(th2)*exp(-(thb - th2)/q);
th = [tha thb];
Ic = [Ica Icb];
Irms = sqrt(1/T*trapz(th, Ic.^2))
```

Conclusion

The code uses a 1024-point data vector for the sample points of the instantaneous capacitor current and the subsequent RMS value integration. The data points are divided proportionately into two vectors in accordance with Equation 2.131. The ripple current computed by the program is 4.84 A.

Exercise 2.18

With Listing 2.14 as a guide, write a MATLAB function that returns the capacitor ripple current from the supplied values of the RMS source voltage, the quality factor, and the load resistance.

Power Computations

The power computations for the HW rectifier with R-C load are illustrated in the next example.

Example 2.19

Compute the average power and related functions for a half-wave rectifier with a 470 μF filter capacitor and a load resistance of 144 Ω. The source voltage is 120 VAC at 60 Hz.

Solution

ON THE CD

The solution code is provided in Listing 2.16 and is available on the CD-ROM as Example2_19.m in folder Chapter 2\Examples and Listings.

LISTING 2.16 Power Computations for the HW Rectifier with R-C Load

```
close all, clear all, clc
Vs = 120;
Vm = Vs*sqrt(2);
f = 60;
T = 2*pi;
w = 2*pi*f;
C = 470E-6;
R = 144;
q = w*R*C;
[th1, th2] = th12hwrc(q);
th = linspace(th1, th2, 1000);
Ith = Vm/R*(q*cos(th) + sin(th));
Is = sqrt(1/T*trapz(th, Ith.^2))
S = Vs*Is
a1 = 2/T*trapz(th, Ith.*cos(th));
b1 = 2/T*trapz(th, Ith.*sin(th));
ph1 = atan(a1/b1);
P = Vm*b1/2
Q = -Vm*a1/2
Is1 = sqrt((a1^2 + b1^2)/2)
D = Vs*sqrt(Is^2 - Is1^2)
```

```
pf = P/S
DPF = cos(ph1)
DF = Is1/Is
```

Conclusion

The code in Listing 2.15 produces the results as follows:

$I_s = 3.70$ A

$S = 444$ VA

$P = 163$ W

$Q = -71$ VAr

$D = 407$ VAd

$pf = 0.37$

$DPF = 0.92$

$DF = 0.40.$

The results reveal that the circuit has a very poor power factor because of the large harmonic content of the source current, indicated by the low distortion factor. The negative VAr reveals that power factor improvement with shunt capacitance is not possible.

EXERCISE 2.19

Compute the diode dissipation and circuit efficiency of the circuit in Example 2.19 if the forward voltage is 0.7 V.

Answers

$P_{DIO} = 0.96$ W and $\eta = 99.4\%$.

FW Rectifier with R-C Load

The FW circuit provides the advantage of a larger DC output voltage than that of the HW circuit. Also, the FW circuit requires less capacitance than the HW circuit for the same specified ripple voltage. The FW rectifier with R-C load is shown in Figure 2.27.

As with the HW circuit with R-C load, the source current of the FW rectifier is periodically interrupted when the source voltage falls below the load voltage. Unlike the HW circuit, however, the diode bridge experiences two conduction intervals for each cycle of the source voltage. The normalized output voltage and source current, along with the rectified source voltage, are illustrated in Figure 2.28.

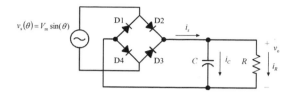

FIGURE 2.27 FW rectifier with R-C load.

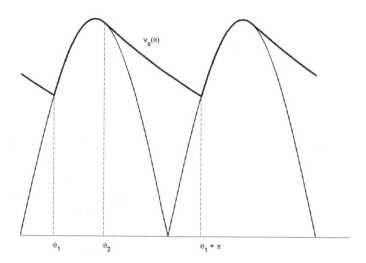

FIGURE 2.28 Output voltage of the FW rectifier with R-C load.

The equations derived for the HW rectifier with R-C load are readily modified for the FW circuit because the FW rectified waveforms have a period of π rather than 2π. In particular, the output voltage for the FW circuit is

$$v_o(\theta) = \begin{cases} V_m \sin\theta, & \theta_1 \leq \theta \leq \theta_2 \\ V_m \sin(\theta_2) e^{-(\theta-\theta_2)/q}, & \theta_2 \leq \theta \leq \theta_1 + \pi \end{cases}, \tag{2.133}$$

and the transcendental equation to determine the cut-in angle is

$$e^{(\theta_2-\pi)/q} \sin\theta_2 = e^{\theta_1/q} \sin\theta_1. \tag{2.134}$$

Furthermore, the average output voltage of the FW circuit is twice that of the HW circuit,

$$V_o = \frac{V_m}{\pi} \left\{ \cos\theta_1 - \cos\theta_2 + q\sin\theta_2 \left[1 - e^{-(\pi-\theta_c)/q} \right] \right\}, \tag{2.135}$$

and the instantaneous capacitor is expressed as

$$i_C(\theta) = \begin{cases} \dfrac{qV_m}{R}\cos\theta, & \theta_1 \le \theta \le \theta_2 \\[3mm] -\dfrac{V_m}{R}\sin(\theta_2)e^{-(\theta-\theta_2)/q}, & \theta_2 \le \theta \le \theta_1 + \pi \end{cases} \tag{2.136}$$

The next example illustrates the average output voltage and output ripple voltage calculations for the FW rectifier with R-C load.

EXAMPLE 2.20

Repeat Exercise 2.16 and Example 2.17 for the FW rectifier with R-C load.

Solution

The solution code is derived from Listing 2.12, with *th2 – 2*pi* changed to *th2 – pi*, and from Listing 2.13 with the average output voltage expressed by Equation 2.133. The code is available as Example 2_20.m in folder Chapter 2\Examples and Listings on the CD-ROM. The output of the program is C = 368 μF, θ_c = 51.7°, V_o = 152 VDC, and V_r = 36.5 V.

ON THE CD

Conclusion

The capacitance for the FW circuit is the same as that of the HW circuit because the quality factor, frequency, and load resistance are the same. The average output voltage, however, is greater for the FW circuit and the ripple voltage is significantly reduced.

EXERCISE 2.20

Write a MATLAB program to compute the value of capacitance for a specified ripple voltage in the FW rectifier with R-C load. Use the program to compute the capacitance required for a ripple voltage of 10 V for the circuit in Example 2.20. What is the average output voltage for this value of capacitance?

Answers

C = 2.5 mF and V_o = 166 VDC.

Power Computations

The diode dissipation in the FW circuit with R-C load is dictated by Equation 2.129 because each diode-pair conducts only once per source cycle. The efficiency of the FW circuit is determined by Equation 2.36 because of the losses in four diodes. The power computations for the FW rectifier with R-C load are illustrated in the next example.

EXAMPLE 2.21

Compute the average power and related functions for the FW rectifier with R-C load with the same circuit parameters as in Example 2.19.

Solution

ON THE CD

The solution code is derived from Listing 2.15, with the period T changed to π, and from Listing 2.12, with the modification applied in Example 2.20. The code is available on the CD-ROM as Example2_21.m in folder Chapter 2\Examples and Listings.

Conclusion

The output of the solution code is as follows:

$I_s = 3.27$ A

$S = 393$ VA

$P = 182$ W

$Q = -53$ VAr

$D = 344$ VAd

$pf = 0.46$

$DPF = 0.96$

$DF = 0.48$.

The FW circuit delivers more power to the load with a smaller RMS source current. Consequently, the apparent power is less than that of the HW circuit, and the power factor is improved. The large VAd content, however, still results in a poor power factor.

EXERCISE 2.21

Repeat Exercise 2.19 for the circuit in Example 2.21.

Answers

$P_{DIO} = 0.4$ W (each) and $\eta = 99.1\%$.

The poor power factor of the previous two circuits is due to the discontinuous source current. If sufficient inductance is introduced into the circuit, the source current becomes continuous and the power factor is improved, with the added benefit of reduced output ripple voltage. The FW rectifier with an inductive-capacitive (*L-C*) filter is considered in the next section.

Full-Wave Rectifier with L-C Filter

The FW rectifier circuit with *L-C* filter is illustrated in Figure 2.29.

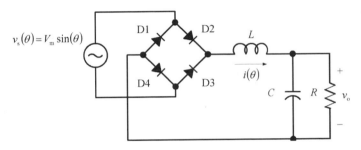

FIGURE 2.29   FW rectifier with *L-C* filter.

If the inductor current is continuous, the rectified source is always connected to the *L-C-R* network. The continuous instantaneous inductor current is plotted in Figure 2.30.

With continuous inductor current, application of KCL and KVL to the circuit of Figure 2.29 yields a set of differential equations in terms of the inductor current and capacitor voltage:

$$V_m \sin\theta = \omega L \frac{di}{d\theta} + v_o. \tag{2.137}$$

$$i = \omega C \frac{dv_o}{d\theta} + \frac{v_o}{R}. \tag{2.138}$$

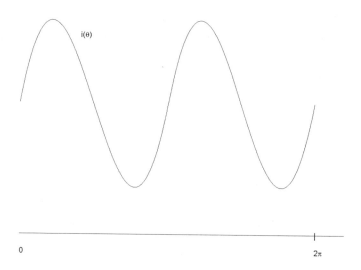

FIGURE 2.30 Continuous instantaneous inductor current.

From Equations 2.137 and 2.138, the differential equation of the inductor current is formulated as

$$\frac{d^2i}{d\theta^2} + \frac{1}{q_c}\frac{di}{d\theta} + \frac{1}{q_c q_i}i = \frac{V_m}{q_i R}\cos\theta + \frac{V_m}{q_c q_i R}\sin\theta, \tag{2.139}$$

in which q_c and q_i are the quality factors of the capacitor and inductor, respectively. The solution to the second-order differential equation has the same form as Equation 2.44. Furthermore, the forced response of the inductor current has the same form as Equation 2.46. However, the development of the complete solution to Equation 2.139 is a very lengthy process. In addition to the derivation of the forced response, the differential equation has boundary conditions that must be applied to solve for the unknown constants in one of three possible natural responses. In lieu of the solution of the differential equation, a Fourier series approach provides a more tractable criterion for continuous inductor current.

Fourier Series Representation of Inductor Current

The Fourier series expansion of the inductor current begins with that of the source voltage. The inductor current harmonics are determined from the ratio of the source voltage harmonics to the impedance of the L-C-R network at each harmonic frequency.

Fourier Series Expansion of the FW Rectified Source Voltage

The even symmetry of the rectified source voltage results in zero sine coefficients in the Fourier series. The cosine coefficients are determined from the integral

$$a_n = \frac{2V_{\mathrm{m}}}{\pi} \int_0^\pi \sin\theta \cos n\theta \, d\theta. \tag{2.140}$$

With the trigonometric identity

$$\sin\alpha\cos\beta = \frac{1}{2}\sin(\alpha+\beta) + \frac{1}{2}\sin(\alpha-\beta), \tag{2.141}$$

the integral in Equation 2.140 becomes two integrals that are readily evaluated:

$$a_n = \frac{V_{\mathrm{m}}}{\pi}\left[\int_0^\pi \sin\big[(1+n)\theta\big]\,d\theta + \int_0^\pi \sin\big[(1-n)\theta\big]\,d\theta \right]. \tag{2.142}$$

The integrals in Equation 2.142 are evaluated as follows:

$$\int_0^\pi \sin\big[(1+n)\theta\big]\,d\theta = \frac{1}{1+n}\Big\{1 - \cos\big[(1+n)\pi\big]\Big\} \tag{2.143}$$

$$\int_0^\pi \sin\big[(1-n)\theta\big]\,d\theta = \frac{1}{1-n}\Big\{1 - \cos\big[(1-n)\pi\big]\Big\}. \tag{2.144}$$

Equations 2.143 and 2.144 are zero for odd values of n. For even values of n,

$$\frac{1}{1+n}\Big\{1 - \cos\big[(1+n)\pi\big]\Big\} = \frac{2}{1+n}, \quad n = 2, 4, 6, \dots. \tag{2.145}$$

$$\frac{1}{1-n}\Big\{1 - \cos\big[(1-n)\pi\big]\Big\} = \frac{2}{1-n}, \quad n = 2, 4, 6 \dots. \tag{2.146}$$

With Equations 2.145 and 2.146, the expression for the cosine coefficients is

$$a_n = \frac{4V_{\mathrm{m}}}{\pi\big(1-n^2\big)}, \quad n = 2, 4, 6, \dots. \tag{2.147}$$

With Equations 2.32 and 2.147, the Fourier series expansion of the FW rectified source voltage is expressed as

$$v_s(\theta) = \frac{2V_m}{\pi} + \frac{4V_m}{\pi} \sum_{n=2,4,\ldots}^{\infty} \left[\frac{1}{1-n^2} \right] \cos n\theta. \tag{2.148}$$

The examples that follow illustrate how to plot the rectified source voltage and its magnitude spectrum with Equation 2.148.

EXAMPLE 2.22

Write a MATLAB program to plot four cycles of an FW rectified sine-wave voltage with a peak value of 170 V.

Solution

ON THE CD

The solution code is provided in Listing 2.17. The code is available on the CD-ROM as file Example2_22.m in folder Chapter 2\Examples and Listings.

LISTING 2.17 Program to Plot Four Cycles of an FW Rectified Sine-Wave

```
close all, clear all, clc
N = 1024;
Vm = 170;
th = linspace(0, pi, 1024);
n = 2:2:40;
vth = 2*Vm/pi + 4*Vm./(pi*(1 - n.^2))*cos(n'*th);
x = linspace(0, 4*pi, 4*N);
Vs = vth'*ones(1, 4);
plot(x, Vs(:)), grid
```

Conclusion

The program in Listing 2.17 computes the even harmonics from 2 to 40. Four cycles of the waveform are stored in matrix **Vs,** which is created from the product of the transpose of **vth** and a row vector of four ones. The colon operator in the plot statement converts all the elements in **Vs** to a column vector so that **x** and **Vs(:)** have the same dimensions. The plot produced by the program is shown in Figure 2.31.

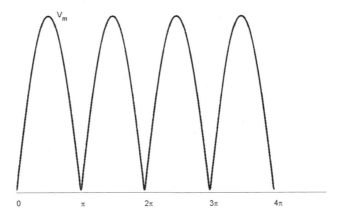

FIGURE 2.31   Four cycles of an FW rectified sine-wave.

EXERCISE 2.22

Modify Listing 2.17 to include harmonics up to and including the 100th harmonic. Plot eight cycles of the waveform.

EXAMPLE 2.23

Plot the discrete magnitude spectrum of the waveform from Example 2.22.

Solution

The solution code appears in Listing 2.18 and is available on the CD-ROM as Example 2_23.m in Chapter 2\Examples and Listings. The discrete spectrum is plotted in Figure 2.32.

LISTING 2.18   Code to Plot the Magnitude Spectrum of an FW Rectified Sine-Wave

```
close all, clear all, clc
Vm = 170;
n = [2:2:40];
Vh = [2*Vm/pi abs(4*Vm/pi*1./(1 - n.^2))];
stem([0 n], Vh, 'k', 'filled'), grid
title('Magnitude Spectrum')
xlabel('harmonic')
```

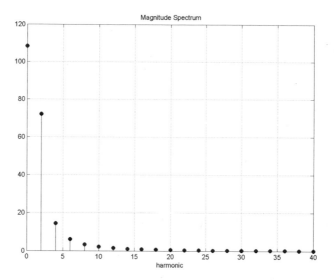

FIGURE 2.32   Magnitude spectrum of the FW rectified sine-wave.

Conclusion

Figure 2.32 reveals that the harmonic magnitudes decrease rapidly after the second harmonic. The magnitudes beyond the eighth harmonic are less than 3% of the DC component.

EXERCISE 2.23

Determine the harmonic number where the RMS values of the harmonics begin to be less than 1% of the DC component.

Answer

The 12th harmonic.

Inductor Current Harmonics

The impedance seen by the rectified source voltage is the parallel combination of the resistance and the capacitance in series with the inductance:

$$Z_s = sL + \frac{\dfrac{R}{sC}}{R + \dfrac{1}{sC}}. \tag{2.149}$$

Substitution of $s = j\omega$ into Equation 2.149 with a rearrangement of the expression over a common denominator results in

$$Z_s = \frac{R\left(1 - \omega^2 LC\right) + j\omega L}{1 + j\omega RC}. \tag{2.150}$$

As stated earlier, the impedance must be computed at each harmonic multiple of the fundamental frequency of the source voltage:

$$Z_s = \frac{R\left(1 - n^2 \omega_0^2 LC\right) + jn\omega_0 L}{1 + jn\omega_0 RC}. \tag{2.151}$$

In terms of the inductive and capacitive quality factors, Equation 2.151 becomes

$$Z_s = \frac{R\left[\left(1 - n^2 q_c q_i\right) + jn q_i\right]}{1 + jn q_c}. \tag{2.152}$$

The inductor is a short circuit to the average value of the rectified source; consequently, the average inductor current is

$$I_0 = \frac{2V_m}{\pi R}. \tag{2.153}$$

With Equations 2.148, 2.152, and 2.153, the Fourier series expansion of the inductor current is

$$i(\theta) = \frac{2V_m}{\pi R} + \frac{4V_m}{\pi |Z_s|} \sum_{n=2,4,\dots}^{\infty} \left[\frac{1}{1 - n^2}\right] \cos\left(n\theta + \angle Z_s\right). \tag{2.154}$$

If the value of inductance is chosen such that the instantaneous inductor current is always positive, then the rectifier bridge is never reverse-biased, the inductor current is continuous, and the rectifier output voltage is dictated by Equation 1.49. The use of Equation 2.154 to determine the inductance numerically is illustrated in the next example.

EXAMPLE 2.24

Determine the inductance required for continuous inductor current for the circuit in Example 2.20 with (a) $C = 100\ \mu F$, (b) $C = 1000\ \mu F$, and (c) $C = 2,500\ \mu F$.

Solution

The solution is provided by the program in Listing 2.19 and is available on the CD-ROM as Example2_24.m in Chapter 2\Examples and Listings.

LISTING 2.19   Program to Compute L for Continuous Inductor Current

```
close all, clear all, clc
tpi = 2*pi;
N = 1024;
f = 60;
w = tpi*f;
Vm = 170;
R = 72;
L = 0.01;
C = 2500E-6;
qc = w*R*C;
n = 2:2:80;
th = linspace(0, pi, N);
inc = 0.01;
Ix = 1;
while abs(Ix) > 1E-6,
    qi = w*L/R;
    Zs = R*((1 - qc*qi*n.^2) + j*n*qi)./(1 + j*n*qc);
    In = 4*Vm./(pi*abs(Zs).*(1 - n.^2));
    Ith = 2*Vm/(pi*R) + In*cos(n'*th + angle(Zs)'*ones(1,N));
    Ix = min(Ith);
    if Ix < 0,
        L = L + inc;
    else
        L = L - inc;
        inc = inc/2;
        L = L + inc;
    end
end
```

Conclusion

The program in Listing 2.19 uses the *min* function to acquire the minimum value of the instantaneous current. The program terminates when the absolute value of the minimum is less than 1E-6. The values of inductance that correspond to the given values of capacitance are (a) $L = 81$ mH, (b) $L = 65$ mH, and (c) $L = 64$ mH.

EXERCISE 2.24

Modify the program in Listing 2.19 to compute the inductance required for a given minimum positive value of inductor current.

Approximate Analysis

The results of Example 2.24 appear to indicate a limiting value of inductance as the value of capacitance increases. As the value of capacitance approaches infinity, the capacitor behaves as a short circuit to all the harmonic content of the inductor current. With infinite capacitance, the impedance seen by the source, Equation 2.149, becomes

$$Z_s = jn\omega_0 L. \tag{2.155}$$

If the inductor current is approximated as only the sum of the average value and the second harmonic, then, with Equation 2.155, the Fourier series in Equation 2.154 simplifies to

$$i(\theta) = \frac{2V_m}{\pi R} + \frac{2V_m}{3\pi\omega_0 L} \sin 2\theta. \tag{2.156}$$

Equation 2.156 is always positive if the magnitude of the second harmonic is less than the average value:

$$\frac{2V_m}{\pi R} > \frac{2V_m}{3\pi\omega_0 L}. \tag{2.157}$$

The solution of the inequality for L is

$$L > \frac{R}{3\omega_0}. \tag{2.158}$$

With Equation 2.158, the value of L for Example 2.24 is computed as

$$L > \frac{1}{5\pi}. \tag{2.159}$$

Equation 2.159 matches the result for L when $C = 2{,}500 \ \mu F$. The approximate analysis thus provides a convenient way to estimate the required inductance simply from the load resistance and the source frequency.

CHAPTER SUMMARY

The diode rectifier circuit is used to convert AC to DC; it creates a waveform with a DC component from a waveform with zero average value. The circuit designer has the option to choose from the economical HW rectifier or the better-performing FW rectifier. Rectifiers are the key subcircuits in DC power supplies and battery chargers that operate from the AC mains. Although the diode or diodes alone produce a DC voltage and current, inductors and capacitors are used to reduce the harmonic content and better approximate a constant output.

The utility of the diode rectifier circuit, however, brings with it issues of power factor and efficiency. The power factor of rectifier circuits are typically poor because of significant harmonic content in the source current. The efficiency of the FW rectifier is significantly affected by the energy dissipation in four diodes.

In spite of these issues, however, the diode rectifier is a viable circuit solution in applications where a relatively fixed DC output voltage or current is required. If a controllable range of output voltage is required, the diode is replaced in the circuit with the SCR. The controlled rectifier circuit is the subject of the next chapter.

ON THE CD

THE MATLAB TOOLBOX

The functions listed below are available in the Toolbox subfolder of Chapter 2 on the CD-ROM.

HWR

Function *hwr.m* computes the average output current of the HW rectifier with resistive load from the RMS source voltage and the load resistance. Also computed are the apparent volt-amperes and average power of the load.

Syntax

```
[Io, S, P] = hwr(Vs, R)
```

FWR

Function *fwr.m* computes the average output current of the FW rectifier with resistive load from the RMS source voltage and the load resistance. Also computed is the average load power.

Syntax

```
[Io, P] = fwr(Vs, R)
```

HWRL

Function *hwrl.m* computes the average output current of the HW rectifier with series *R-L* impedance from the RMS source voltage, the source frequency, and the load inductance and resistance. Also computed are the apparent, real, reactive, and distortive powers along with the displacement, distortion, and power factors.

Syntax

```
[Io, S, P, Q, D, DPF, DF, pf] = hwrl(Vs, f, L, R)
```

FWRL

Function *fwrl.m* computes the average output current of the FW rectifier with series *R-L* impedance from the RMS source voltage the source frequency, and the load inductance and resistance. Also computed are the apparent, real, reactive, and distortive powers along with the displacement, distortion, and power factors.

Syntax

```
[Io, S, P, Q, D, DPF, DF, pf] = fwrl(Vs, f, L, R)
```

HWRV

Function *hwrv.m* computes the average output current of the HW charger from the RMS source voltage, the source frequency, the series resistance, and the battery voltage. Also computed are the apparent, real, reactive, and distortive powers along with the displacement, distortion, and power factors.

Syntax

```
[Io, S, P, Q, D, DPF, DF, pf] = hwrv(Vs, f, R, Vb)
```

FWRV

Function *fwrv.m* computes the average output current of the FW charger from the RMS source voltage, the source frequency, the series resistance, and the battery voltage. Also computed are the apparent, real, reactive, and distortive powers along with the displacement, distortion, and power factors.

Syntax

```
[Io, S, P, Q, D, DPF, DF, pf] = fwrv(Vs, f, R, Vb)
```

HWRLV

Function *hwrlv.m* computes the average output current of the HW charger with series R-L impedance from the RMS source voltage, the source frequency, the series inductance and resistance, and the battery voltage. Also computed are the apparent, real, reactive, and distortive powers along with the displacement, distortion, and power factors.

Syntax

```
[Io, S, P, Q, D, DPF, DF, pf] = hwrlv(Vs, f, L, R, Vb)
```

FWRLV

Function *fwrlv.m* computes the average output current of the FW charger with series R-L impedance from the RMS source voltage, the source frequency, the series inductance and resistance, and the battery voltage. Also computed are the apparent, real, reactive, and distortive powers along with the displacement, distortion, and power factors.

Syntax

```
[Io, S, P, Q, D, DPF, DF, pf] = fwrlv(Vs, f, L, R, Vb)
```

HWRC

Function *hwrc.m* computes the average output voltage of the HW rectifier with R-C load from the RMS source voltage, the source frequency, and the load resistance and capacitance. Also computed are the apparent, real, reactive, and distortive powers along with the displacement, distortion, and power factors.

Syntax

```
[Vo, S, P, Q, D, DPF, DF, pf] = hwrc(Vs, f, R, C)
```

FWRC

Function *fwrc.m* computes the average output voltage of the FW rectifier with R-C load from the RMS source voltage, the source frequency, and the load resistance and capacitance. Also computed are the apparent, real, reactive, and distortive powers along with the displacement, distortion, and power factors.

Syntax

```
[Vo, S, P, Q, D, DPF, DF, pf] = fwrc(Vs, f, R, C)
```

PROBLEMS

See Chapter 7 for a review of transformer fundamentals.

Problem 1

Specify a transformer (turns-ratio and kVA rating) for a HW rectifier that supplies 1000 W to a 10 Ω-load resistance from a 120 VAC source.

Problem 2

Specify a transformer for an FW rectifier that supplies 1000 W to a 10 Ω load resistance from a 120 VAC source.

Problem 3

An HW rectifier with R-L load absorbs 500 W from a 120 VAC, 60 Hz source. With a load resistance of 10 Ω, determine the inductance required to limit the average load current to 5 A.

Problem 4

An FW rectifier with R-L load absorbs 500 W from a 120 VAC, 60 Hz source. With a load resistance of 10 Ω, determine the inductance required to limit the average load current to 5 A.

Problem 5

Specify a transformer for an HW battery charger to supply an average current of 10 A to a 12 V battery with a series resistance of 0.2 Ω from a 120 VAC source. Compute the efficiency of the circuit if $V_f = 1.4$ V.

Problem 6

Specify a transformer for an FW battery charger to supply an average current of 10 A to a 12 V battery with a series resistance of 0.2 Ω from a 120 VAC source. Compute the efficiency of the circuit if $V_f = 1.4$ V.

Problem 7

Specify a transformer for an HW battery charger to supply an average current of 10 A to a 12 V battery with a series impedance of $0.2 + j0.3$ Ω from a 120 VAC, 60 Hz source. Compute the efficiency of the circuit if $V_f = 1.4$ V.

Problem 8

Specify a transformer for an FW battery charger to supply an average current of 10 A to a 12 V battery with a series impedance of 0.2 + j0.3 Ω from a 120 VAC, 60 Hz source. Compute the efficiency of the circuit if V_f = 1.4 V.

Problem 9

Specify a capacitor (capacitance and ripple current) for an HW rectifier to supply 48 VDC with 10% ripple to a 7.5 Ω load. Specify the supply transformer to the circuit for a 120 VAC, 60 Hz source.

Problem 10

Specify a capacitor (capacitance and ripple current) for an FW rectifier to supply 48 VDC with 10% ripple to a 7.5 Ω load. Specify the supply transformer to the circuit for a 120 VAC, 60 Hz source.

Problem 11

Compute the efficiency of the circuit in Problem 10 if V_f = 1.2 V.

Problem 12

Calculate the series inductance required for the circuit in Problem 10 to draw continuous current from the source.

3 Phase-Controlled Rectifier Circuits

In This Chapter

- Introduction
- Phase-Controlled Rectification
- Phase-Controlled Chargers
- Chapter Summary
- The MATLAB Toolbox
- Problems

INTRODUCTION

Chapter 2 presented the diode as an uncontrolled rectifier because the conduction time of the device is determined exclusively by the source voltage. This chapter deals with the silicon-controlled rectifier, also called a *thyristor,* which has a third terminal in addition to an anode and a cathode. This *gate* terminal controls the conduction time of the device along with the polarity of the source voltage.

The purpose of the controlled rectifier circuit is the same as that of the diode rectifier: to produce a DC voltage and current from an AC source. The controlled rectifier, however, offers the advantage of a variable DC output.

The method of analysis of the controlled rectifier is the same as that followed in Chapter 2: formulation of the instantaneous source current followed by the derivations of the expressions for the average load voltage, load current, and thyristor current. Computations of the apparent, real, reactive, and distortive powers are also demonstrated along with those of the power, displacement, and distortion factors. Power factor improvement with shunt capacitance is also considered.

Since the source current conduction interval is controlled by the gate terminal, all the derivations of the circuit equations are a function of the *delay angle*—the angle relative to the source voltage at which the thyristor begins to conduct current.

PHASE-CONTROLLED RECTIFICATION

The SCR is a three-terminal device. It has an anode, a cathode, and a gate. The schematic symbol for the SCR is illustrated in Figure 3.1, in which the letter k is used to indicate the cathode. The anode-to-cathode voltage V_{ak} is similar to the forward voltage of a diode.

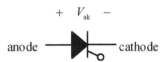

FIGURE 3.1   Silicon-controlled rectifier.

As in the analysis of diode rectifiers, the voltage across the SCR is often neglected because the voltages in the circuit are of much greater magnitude. When SCR power dissipation is considered, however, V_{ak} cannot be assumed to be zero.

As the diode, the SCR only conducts current from anode to cathode and only when forward-biased. However, in addition to being forward-biased, a small current must be injected into the gate terminal to cause current conduction from anode to cathode. This *trigger current* need not be present after the SCR begins to conduct; only a short burst of gate current is required to trigger the device into conduction. The duration and magnitude of the trigger current is specified by the SCR manufacturer.

Once the SCR begins to conduct, it will continue to conduct until the anode current is extinguished; gate-control of the device is completely lost after conduction begins. Once the anode current is zero, the SCR returns to the high-impedance *blocking state*. The transition of the SCR from the conduction state to the blocking state is called *commutation*. An alternate way to commutate the SCR is to reverse-bias the device for an amount of time equal to the turn-off time, t_q, also known as the *quench* time. The SCR is thus well suited for applications in which the source voltage reverses polarity, that is, AC circuit applications. In an AC circuit the SCR is said to be *naturally commutated* by the source. The circuit that provides gate current from an AC source is synchronized with the source voltage waveform to initiate conduction at the same point in time of each source cycle. This method of gate current synchronization is called *phase control*.

HW Phase-Controlled Rectifier with Resistive Load

The simplest phase-controlled rectifier is the HW circuit with resistive load shown in Figure 3.2.

The load voltage and current waveforms of the circuit, along with the gate trigger current, are illustrated in Figure 3.3.

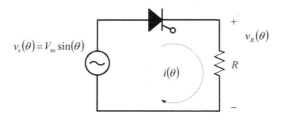

FIGURE 3.2 HW phase-controlled rectifier with resistive load.

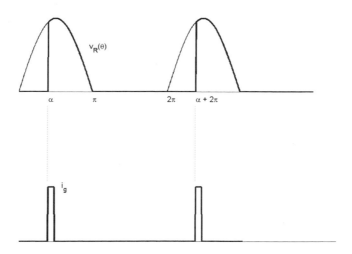

FIGURE 3.3 Load voltage and current waveforms of the HW phase-controlled rectifier.

As indicated in Figure 3.3, the SCR is in the blocking state until the gate is triggered at angle α. Once triggered, the SCR continues to conduct until the source voltage reverses polarity and the anode current is extinguished. On the assumption of negligible V_{ak}, the instantaneous load voltage and current are expressed by

$$v_R(\theta) = \begin{cases} V_m \sin\theta, & \alpha \leq \theta \leq \pi \\ 0, & \pi < \theta \leq 2\pi \end{cases}. \tag{3.1}$$

$$i(\theta) = \begin{cases} \dfrac{V_m}{R} \sin\theta, & \alpha \leq \theta \leq \pi \\ 0, & \pi < \theta \leq 2\pi \end{cases}. \tag{3.2}$$

Average Load Voltage and Current

The average load voltage is determined from

$$V_o = \frac{1}{2\pi} \int_\alpha^\pi V_m \sin\theta \, d\theta. \tag{3.3}$$

Evaluation of the integral in Equation 3.3 results in

$$V_o = \frac{V_m}{2\pi}\left(1 + \cos\alpha\right). \tag{3.4}$$

If $\alpha = 0$ (no delay), then Equation 3.4 reverts to Equation 2.4 from Chapter 2. As α approaches π, the average voltage approaches zero. The average current is the average voltage divided by the load resistance

$$I_o = \frac{V_m}{2\pi R}\left(1 + \cos\alpha\right). \tag{3.5}$$

Example 3.1

A half-wave phase-controlled rectifier is supplied by a 120 VAC source voltage. Determine the required delay angle for a DC output voltage of 50 VDC.

Solution

The solution of Equation 3.4 for the delay angle is

$$\alpha = \arccos\left(\frac{2\pi V_{\mathrm{o}}}{V_{\mathrm{m}}} - 1\right). \tag{3.6}$$

Substitution of the respective voltages into Equation 3.6 yields a delay angle of 31.7°.

Conclusion

The required delay is readily determined from Equation 3.6 for any DC output voltage in the range of $0 \le V_{\mathrm{o}} \le \dfrac{V_{\mathrm{m}}}{\pi}$.

EXERCISE 3.1

Determine the delay angle for the circuit in Example 3.1 that results in a DC current of 10 A with a load resistance of 2 Ω.

Answer

$\alpha = 105°.$

Average Power and Efficiency

The RMS source current is determined from Equation 3.2:

$$I_{\mathrm{RMS}} = \sqrt{\frac{1}{2\pi} \int_{\alpha}^{\pi} \frac{V_{\mathrm{m}}^2}{R^2} \sin^2 \theta \, d\theta}. \tag{3.7}$$

Evaluation of the integral in Equation 3.7 results in the RMS value expression

$$I_{\mathrm{RMS}} = \frac{V_{\mathrm{m}}}{2R} \sqrt{1 - \frac{\alpha}{\pi} + \frac{1}{2\pi} \sin 2\alpha}. \tag{3.8}$$

With Equation 3.8, the apparent power of the source is expressed as

$$S = \frac{V_{\mathrm{m}}^2}{2R} \sqrt{\frac{1}{2} - \frac{\alpha}{2\pi} + \frac{1}{4\pi} \sin 2\alpha}, \tag{3.9}$$

and the average power absorbed by the load is

$$P = \frac{V_m^2}{4R}\left[1 - \frac{\alpha}{\pi} + \frac{1}{2\pi}\sin 2\alpha\right]. \tag{3.10}$$

The power factor is the ratio of Equation 3.10 to Equation 3.9:

$$pf = \sqrt{\frac{1}{2} - \frac{\alpha}{2\pi} + \frac{1}{4\pi}\sin 2\alpha}. \tag{3.11}$$

When $\alpha = 0$, Equation 3.11 reverts to Equation 2.13 in Chapter 2. The next example illustrates the effect of the delay angle on the average power and power factor of the HW phase-controlled rectifier.

EXAMPLE 3.2

Write a MATLAB program to plot the normalized average power of the HW phase-controlled rectifier along with the power factor.

Solution

ON THE CD

The solution is provided by the program in Listing 3.1. The program is also available on the CD-ROM as file Example3_2.m in folder Chapter 3\Examples and Listings. The normalized power and power factor functions are plotted in Figure 3.4.

LISTING 3.1   Program to Plot Normalized Average Power and Power Factor

```
close all, clear all, clc
tpi = 2*pi;
a = linspace(0, pi, 1024);
pf = sqrt(0.5 - a/tpi + 0.5/tpi*sin(2*a));
Pn = 0.5*pf.^2;
plot(a, Pn, 'k', a, pf, 'k')
xlabel('delay angle (radians)')
text(1.5, 0.6, 'Power Factor')
text(0.5, 0.3, 'Normalized Power')
```

Conclusion

With a delay angle of zero, the normalized average power and power factor revert to those of the HW diode rectifier with resistive load. The functions are monotonically decreasing; there is no delay angle other than zero that offers maximum power or a maximum power factor. The power factor only continues to erode as the delay angle increases.

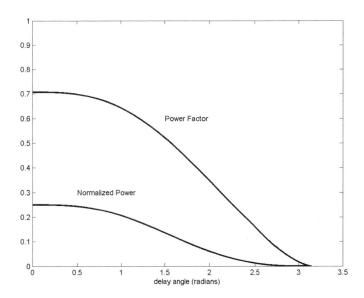

FIGURE 3.4 Normalized average power and power factor of the HW phase-controlled rectifier.

EXERCISE 3.2

A half-wave phase-controlled rectifier is supplied by a 120 VAC source and has a load resistance of 25 Ω. Find the delay angle that produces a load-absorbed power of 100 W.

Answer

$\alpha = 136°$.

Efficiency

In the previous equations V_{ak} was assumed negligible to facilitate the analyses. In a practical circuit the SCR power dissipation is the product of V_{ak} and the average load current:

$$P_{SCR} = \frac{V_{ak} V_m}{2\pi R}(1 + \cos\alpha).$$ (3.12)

The total power delivered by the source to the circuit is the sum of the output power and the power dissipation in the SCR. The efficiency, therefore, is

$$\eta = \frac{P_o}{P_o + P_{SCR}}.$$ (3.13)

Power Computations

As stated earlier and demonstrated by the equations derived thus far, the delay angle appears in all the circuit functions. In lieu of explicit derivations, the numerical approach taken in Chapter 2 provides more tractable solutions than their algebraic counterparts. The next example illustrates the power computations for the HW phase-controlled rectifier.

EXAMPLE 3.3

A half-wave phase-controlled rectifier is supplied by a 120 VAC, 60 Hz source and has a load resistance of 10 Ω. Compute the apparent, real, reactive, and distortive powers, along with the power factor, displacement, and distortion factors for a delay angle of 30°. Compute the SCR dissipation for $V_{ak} = 1.7$ V and the circuit efficiency. Discuss the possibility of power factor improvement with shunt capacitance.

Solution

ON THE CD

The solution code is provided in Listing 3.2 and is available as Example3_3.m on the CD-ROM in folder Chapter 3\Examples and Listings.

LISTING 3.2   Program to Perform Power Computations for the HW Phase-Controlled Rectifier

```
close all, clear all, clc
tpi = 2*pi;
Vak = 1.7;
Vs = 120;
Vm = Vs*sqrt(2);
R = 10;
a = 30;
a = a*pi/180;
th = linspace(a, pi, 1024);
Ith = Vm/R*sin(th);
Is = sqrt(1/tpi*trapz(th, Ith.^2));
a1 = 2/tpi*trapz(th, Ith.*cos(th));
b1 = 2/tpi*trapz(th, Ith.*sin(th));
Is1 = sqrt((a1^2 + b1^2)/2);
S = Vs*Is
P = Vm*b1/2
Q = -Vm*a1/2
D = Vs*sqrt(Is^2 - Is1^2)
pf = P/S
ph1 = atan(a1/b1);
DPF = cos(ph1)
DF = Is1/Is
```

```
Pscr = Vak*Vm/(2*pi*R)*(1 + cos(a))
eta = P/(P + Pscr)
```

The results produced by the program are as follows:

$S = 1.00$ kVA

$P = 699$ W

$Q = 57.3$ VAr

$D = 717$ VAd

$pf = 0.70$

$DPF = 1.00$

$DF = 0.70$

$P_{SCR} = 8.57$ W

$\eta = 98.8\%$.

Conclusion

The results are typical for a HW rectifier circuit. A delay angle of 30° is only a 17% reduction of the conduction angle of an HW diode rectifier. Power factor improvement is possible but not very effective because the displacement factor is near unity. In addition, the value of shunt capacitance varies with the delay angle. The circuit is very efficient, but the dissipation in the SCR is significant.

EXERCISE 3.3

Use the program in Listing 3.1 to obtain data points of the reactive power as the delay angle is varied from 0 to 180° in increments of 10°. Plot the reactive power versus delay angle.

FW Phase-Controlled Rectifier with resistive load

Two circuit implementations of the FW phase-controlled rectifier with resistive load are illustrated in Figure 3.5.

The circuit in Figure 3.5a requires a more complicated gate-trigger circuit than the one in Figure 3.5b but is more efficient because it uses one less semiconductor device. The load voltage waveform of the FW phase-controlled rectifier circuit is plotted in Figure 3.6.

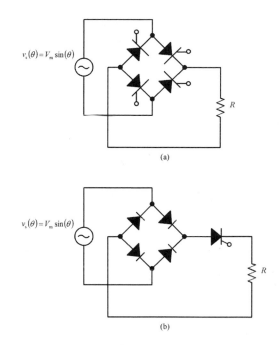

(a)

(b)

FIGURE 3.5 FW phase-controlled rectifier circuits.

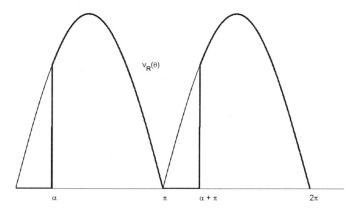

FIGURE 3.6 Load voltage waveform of the FW phase-controlled rectifier circuit.

A comparison of the waveforms in Figure 3.6 with those in Figure 3.3 reveals a similarity for which the only difference is the period: 2π for the HW circuit and π for the FW circuit. Thus, the relevant equations for the FW phase-controlled rectifier are

$$V_o = \frac{V_m}{\pi}(1 + \cos\alpha), \tag{3.14}$$

$$I_o = \frac{V_m}{\pi R}(1 + \cos\alpha), \tag{3.15}$$

$$I_{RMS} = \frac{V_m}{R}\sqrt{\frac{1}{2} - \frac{\alpha}{2\pi} + \frac{1}{4\pi}\sin 2\alpha}, \tag{3.16}$$

$$S = \frac{V_m^2}{R}\sqrt{\frac{1}{4} - \frac{\alpha}{4\pi} + \frac{1}{8\pi}\sin 2\alpha}, \tag{3.17}$$

$$P = \frac{V_m^2}{R}\left[\frac{1}{2} - \frac{\alpha}{2\pi} + \frac{1}{4\pi}\sin 2\alpha\right], \tag{3.18}$$

and

$$pf = \sqrt{1 - \frac{\alpha}{\pi} + \frac{1}{2\pi}\sin 2\alpha}. \tag{3.19}$$

The power dissipation in the thyristors of the circuit shown in Figure 3.5a is also dictated by Equation 3.12 because each pair of devices conducts only once for each period of the source. The total device dissipation in the circuit is four times the dissipation of one device. Thus, the circuit efficiency is

$$\eta = \frac{P_o}{P_o + 4P_{SCR}}. \tag{3.20}$$

The SCR in the circuit shown in Figure 3.5b conducts twice per source period; the power dissipation in the thyristor is therefore twice that expressed by Equation 3.12:

$$P_{SCR} = \frac{V_{ak}V_m}{\pi R}(1 + \cos\alpha). \tag{3.21}$$

The diode currents of the circuit shown in Figure 3.5b comprise the thyristor current with a 50% duty ratio. The diode dissipation is thus similar to Equation 3.12 but with V_{ak} replaced with V_f:

$$P_{DIO} = \frac{V_f V_m}{2\pi R}(1+\cos\alpha). \qquad (3.22)$$

The power supplied by the source is the sum of the output power, the total diode dissipation, and the SCR. The circuit efficiency is thus

$$\eta = \frac{P_o}{P_o + 4P_{DIO} + P_{SCR}}. \qquad (3.23)$$

EXAMPLE 3.4

Modify the program in Listing 3.2 to perform the power computations for the FW phase-controlled rectifier with the same parameters as the circuit in Example 3.3. Compare the efficiencies of the two circuits shown in Figure 3.5 if the diode forward voltage is $V_f = 0.9$ V.

Solution

ON THE CD

The solution is provided by program file Example3_4.m in Chapter 3\Examples and Listings on the CD-ROM. The results of the program are as follows:

$S = 1.42$ kVA

$P = 1.40$ kW

$Q = 115$ VAr

$D = 212$ VAd

$pf = 0.99$

$DPF = 1.00$

$DF = 0.99$

$\eta = 97.6\%$ (Figure 3.5a)

$\eta = 97.5\%$ (Figure 3.5b).

Conclusion

As expected, the real and reactive powers are twice those of the HW circuit. The distortive power, however, is greatly reduced, and the power factor is much improved. There is little difference between the efficiencies of the two circuits in Figure 3.5. The decision of which circuit to use in practice is influenced by delay angle control and thermal management issues.

EXERCISE 3.4

Repeat Exercise 3.3 for the circuit in Example 3.4.

HW Phase-Controlled Rectifier with R-L Load

As stated earlier in Chapter 2, purely resistive electrical loads are rare. A more practical load model includes series inductance. The HW phase-controlled rectifier with R-L load is shown in Figure 3.7.

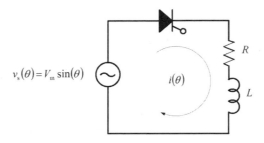

$$v_s(\theta) = V_m \sin(\theta)$$
$$i(\theta)$$

FIGURE 3.7 HW phase-controlled rectifier with R-L load.

The instantaneous current of the circuit in Figure 3.7 has the same form as Equation 2.56 in Chapter 2. The initial condition for the HW phase-controlled controlled rectifier is

$$i(\alpha) = 0. \tag{3.24}$$

Substitution of the initial condition into Equation 2.56 yields the expression

$$i(\theta) = \frac{V_m}{R} \cos\phi \left[\sin(\theta - \phi) - \sin(\alpha - \phi)e^{-(\theta-\alpha)/q} \right], \quad \alpha \le \theta \le \beta. \tag{3.25}$$

Because of the energy stored in the inductance, the current continues to flow after the source voltage has become negative. The anode current is finally extinguished when $\theta = \beta$. A plot of the instantaneous load current is shown in Figure 3.8. Substitution of $\theta = \beta$ into Equation 3.25 yields the transcendental equation

$$e^{\alpha/q} \sin(\alpha - \phi) = e^{\beta/q} \sin(\beta - \phi), \tag{3.26}$$

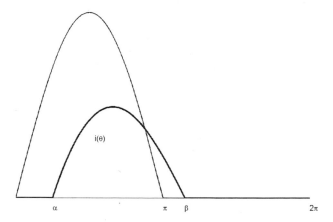

FIGURE 3.8 Load current waveform of the HW phase-controlled rectifier with R-L load.

with which the extinction angle is determined. Figure 3.8 indicates that the solution interval for the extinction angle is in the range $\pi < \beta < 2\pi$. A numerical solution to Equation 3.26 is provided by the MATLAB function *betasolv* on the CD-ROM as file betasolv.m in the Chapter 3\Toolbox folder.

ON THE CD

Average Load Current

Application of the average value integral to Equation 3.25 results in

$$I_o = \frac{V_m}{2\pi R} \cos\phi \int_\alpha^\beta \left[\sin(\theta - \phi) - \sin(\alpha - \phi)e^{-(\theta-\alpha)/q} \right] d\theta. \tag{3.27}$$

Evaluation of the integral in Equation 3.27 yields the average load current expression

$$I_o = \frac{V_m}{2\pi R} \cos\phi \left[\cos(\alpha - \phi) - \cos(\beta - \phi) - q\sin(\alpha - \phi)\left(1 - e^{-\theta_c/q}\right) \right], \tag{3.28}$$

in which $\theta_c = \beta - \alpha$ is the current conduction angle. The power absorbed in the thyristor is the product of the average current and the anode-to-cathode voltage drop:

$$P_{SCR} = V_{ak} I_o. \tag{3.29}$$

Since there is only one SCR in the circuit, the efficiency is dictated by Equation 3.11.

Power Computations

The average power and power factor computations for the HW phase-controlled rectifier with R-L load are illustrated in the next example.

EXAMPLE 3.5

Compute the RMS source current, average output current, apparent power, power absorbed in the load, and the power factor of the circuit in Figure 3.8 with a 220 VAC source voltage, a load impedance of $30 + j20\Omega$ (measured at the source frequency), and a delay angle of 45°.

Solution

ON THE CD

The solution code is provided in Listing 3.3 and is also available on the CD-ROM as Example3_5.m in the Chapter 3\Examples and Listings folder.

LISTING 3.3 Solution Program for Example 3.5

```
close all, clear all, clc
T = 2*pi;
Vs = 220;
Vm = Vs*sqrt(2);
R = 30;
X = 20;
q = X/R;
phi = atan(q);
a = 45;
a = a*pi/180;
B = betasolv(a, q);
thc = B - a;
th = linspace(a, B, 1024);
Ith = Vm/R*cos(phi)*(sin(th - phi) - sin(a - phi)*exp(-(th - a)/q));
Is = sqrt(1/T*trapz(th, Ith.^2));
ap = a - phi;
bp = B - phi;
Io = Vm/(2*pi*R)*cos(phi)*(cos(ap) - cos(bp) - q*sin(ap)*...
    (1 - exp(-thc/q)));
S = Vs*Is
P = (Is^2)*R
pf = P/S
```

Conclusion

The results of the program are as follows:

$I_{RMS} = 4.11$ A

$I_o = 2.54$ A

$S = 904$ VA

$P = 507$ W

$pf = 0.56.$

The impedance of a reactive load is often measured at the source frequency. The quality factor is readily determined from the measured impedance as the ratio of the inductive reactance to the resistance. The program makes use of the *betasolv* function to determine the extinction angle. The poor factor of the circuit arises from the relatively large RMS source current in comparison with the average load current. The larger RMS source current is indicative of significant harmonic content.

EXERCISE 3.5

Compute the reactive and distortive powers of the circuit in Example 3.5 along with the displacement and distortion factors.

Answers

$Q = 388$

$D = 641$

$DPF = 0.79$

$DF = 0.71.$

The HW phase-controlled rectifier is an economical circuit because only one SCR is required with its associated control circuitry. The power factor, however, is quite poor, and power factor improvement with shunt capacitance requires a capacitance value that is dependent upon the delay angle. In situations where the power factor is an issue, the FW phase-controlled rectifier is a viable circuit solution.

FW Phase-Controlled Rectifier with R-L Load

The FW phase-controlled rectifier with R-L load is modeled by the schematic shown in Figure 3.5a with an inductance placed in series with the resistive load. The complete circuit is illustrated in Figure 3.9.

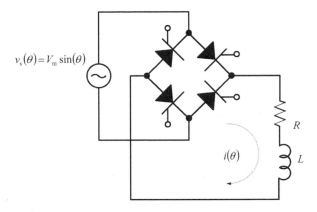

FIGURE 3.9   FW phase-controlled rectifier circuit with R-L load.

Unlike the HW circuit, the four-thyristor circuit has two modes of operation: discontinuous load current and continuous load current.

Discontinuous Load Current

When the load current is discontinuous, the instantaneous current is the same as that expressed by Equation 3.25 but with a period of π rather than 2π. The discontinuous load current is plotted in Figure 3.10 along with the rectified source voltage.

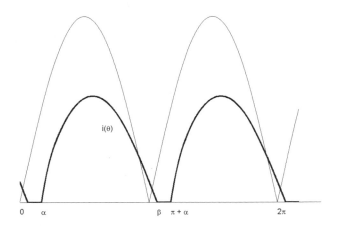

FIGURE 3.10   Discontinuous current of the FW phase-controlled rectifier.

Since the period of the FW circuit is half that of the HW circuit, the average load current is twice that expressed by Equation 3.28:

$$I_\text{o} = \frac{V_\text{m}}{\pi R} \cos\phi \left[\cos(\alpha - \phi) - \cos(\beta - \phi) - q\sin(\alpha - \phi)\left(1 - e^{-\theta_c/q}\right) \right]. \tag{3.30}$$

Continuous Load Current

When the load current is continuous, Equation 2.56 in Chapter 2 is subject to the boundary condition

$$i(\alpha) = i(\alpha + \pi). \tag{3.31}$$

Application of the boundary condition to Equation 2.56 yields

$$A_0 e^{-\alpha/q} + \frac{V_\text{m}}{R}\cos\phi\sin(\alpha - \phi) = A_0 e^{-(\alpha+\pi)/q} - \frac{V_\text{m}}{R}\cos\phi\sin(\alpha - \phi). \tag{3.32}$$

The solution of Equation 3.32 for the unknown constant A_0 is

$$A_0 = -\frac{2V_\text{m} e^{-\alpha/q}}{R(1 - e^{-\pi/q})}\cos\phi\sin(\alpha - \phi). \tag{3.33}$$

Substitution of Equation 3.33 into Equation 2.56 yields the continuous current expression of the FW phase-controlled rectifier with R-L load:

$$i(\theta) = \frac{V_\text{m}}{R}\cos\phi \left[\sin(\theta - \phi) - \sin(\alpha - \phi)\,\text{csch}\left(\frac{\pi}{2q}\right) e^{-(\theta - \chi)/q} \right]. \tag{3.34}$$

In Equation 3.34 *csch* is the hyperbolic cosecant, and $\chi = \alpha + \pi/2$. The continuous load current expression is plotted in Figure 3.11.

A comparison of Figures 3.10 and 3.11 reveals the boundary between the discontinuous and continuous modes of operation:

$$\beta < \alpha + \pi. \tag{3.35}$$

The load current is thus discontinuous if the extinction angle is less than the delay angle at the start of the next period. Stated differently, the current is continuous if the value of delay angle yields a value of current greater than zero at $\theta = \alpha + \pi$.

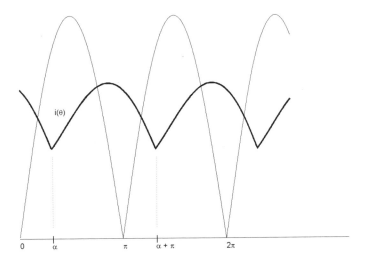

FIGURE 3.11   Continuous current of the FW phase-controlled rectifier with R-L load.

Thus, the boundary between continuous and discontinuous operation occurs at the solution of Equation 3.34 for α when the current is zero and $\theta = \alpha + \pi$:

$$\sin(\alpha + \pi - \phi) - \sin(\alpha - \phi)\operatorname{csch}\left(\frac{\pi}{2q}\right)e^{-\pi/2q} = 0. \tag{3.36}$$

Equation 3.36 simplifies to

$$\sin(\phi - \alpha)\left[1 + \operatorname{csch}\left(\frac{\pi}{2q}\right)e^{-\pi/2q}\right] = 0, \tag{3.37}$$

which has the solution $\alpha = \phi$. In order to have positive, and therefore continuous, current, α must be less than ϕ:

$$\alpha < \phi. \tag{3.38}$$

EXAMPLE 3.6

A FW phase-controlled rectifier with a load impedance of $Z = 10 + j20\ \Omega$ operates with a delay angle of $\alpha = 30°$. Determine whether the load current is continuous or discontinuous. Determine the range of α that results in continuous current operation.

Solution

The q of the circuit is $X/R = 20/10 = 2$. With the *betasolv* function, the extinction angle is computed to be $\beta = 4.33$ radians, which is greater than $\alpha + \pi = 3.67$ radians. The current is therefore continuous. Also, angle ϕ (63°) is greater than α. Therefore, the range of α that results in continuous current operation is $0 \le \alpha \le 63°$.

Conclusion

It is the smaller delay angles that result in continuous current operation. Smaller delay angles allow more energy to be stored in the inductance, which produces a longer conduction angle and results in continuous current.

EXERCISE 3.6

Determine the range of delay angles for continuous and discontinuous current for a full-wave phase-controlled rectifier with a load impedance of $Z = 7 + j5\ \Omega$.

Answers

The continuous current range is $0 \le \alpha < 36°$; the discontinuous current range is $36° < \alpha < 180°$.

Power Computations

The power computations for the FW circuit are readily computed with a period change from 2π to π. The efficiency of the circuit is expressed by Equation 3.20, in which the power dissipated in the thyristor is computed with Equations 3.28 and 3.29. The next example provides a comparison of the power computations of the FW and HW phase-controlled rectifiers with R-L load.

EXAMPLE 3.7

Repeat the computations from Example 3.5 for a full-wave circuit with the same circuit parameters stated in the example. Compare the results of the FW and HW computations.

Solution

ON THE CD
The solution program is available on the CD-ROM as Example3_7.m in folder Chapter 3\Examples and Listings. The results produced by the program are $I_o = 5.1$ A, $P = 1014$ W, and $pf = 0.79$.

Conclusion

The FW circuit outputs twice the average current and twice the average power as well. The power factor is also significantly improved.

EXERCISE 3.7

Compute the reactive and distortive volt-amperes of the circuit in Example 3.7 along with the displacement and distortion factors.

Answers

$Q = 776$ VAr

$D = 78$ VAd

$DPF = 0.79$

$DF = 1.00.$

PHASE-CONTROLLED CHARGERS

A very common application of the phase-controlled rectifier is in battery charging applications. Adjustment of the thyristor delay angle provides precise control over the charge delivered to the battery. In the circuit analyses that follow, the battery is modeled as an ideal DC voltage source in series with resistive and resistive-inductive (*R-L*) elements.

HW Phase-Controlled Charger with Series Resistance

The HW phase-controlled charger with series resistance is shown in Figure 3.12. In the figure, the load is modeled as an independent DC voltage source with internal resistance.

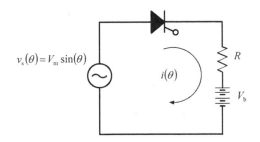

FIGURE 3.12 HW phase-controlled charger with series resistance.

The presence of the battery limits the minimum delay angle of the circuit because the thyristor is not forward-biased until the source voltage exceeds the battery voltage. Furthermore, the conduction angle of the SCR is limited because the thyristor becomes reverse-biased when the source voltage falls below the battery voltage. The instantaneous battery current, along with the battery voltage, is plotted in Figure 3.13.

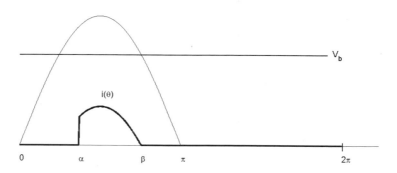

FIGURE 3.13   Instantaneous battery current of the HW phase-controlled charger.

Although the thyristor does not cease conduction immediately when reverse-biased, the quench time is insignificant compared to the period of a typical SCR application.

During conduction of the SCR, the current through the circuit is

$$i(\theta) = \frac{V_m}{R}\left(\sin\theta - \gamma\right), \ \alpha < \theta < \beta, \tag{3.39}$$

in which γ is the ratio of the battery voltage to the peak source voltage,

$$\gamma = \frac{V_b}{V_m}, \tag{3.40}$$

and β is the angle at which the thyristor is reverse-biased:

$$\beta = \pi - \arcsin(\gamma). \tag{3.41}$$

Circuit control is thus limited to a delay angle range of

$$\alpha_{min} < \alpha < \beta, \tag{3.42}$$

in which the minimum delay angle is

$$\alpha_{min} = \arcsin(\gamma). \tag{3.43}$$

The HW phase-controlled charger is much like the HW diode charger in Chapter 2, with the exception that the forward-bias and reverse-bias angles are not symmetrical in the phase-controlled circuit.

Average Battery Current

The average current supplied to the battery is determined from

$$I_b = \frac{V_m}{2\pi R} \int_\alpha^\beta (\sin\theta - \gamma) d\theta. \tag{3.44}$$

Evaluation of the integral in Equation 3.44 yields the expression

$$I_b = \frac{V_m}{2\pi R} \Big[\cos\alpha - \cos\beta - \gamma(\beta - \alpha)\Big]. \tag{3.45}$$

The next example illustrates an application of Equation 3.45.

EXAMPLE 3.8

Determine the delay angle of a half-wave phase-controlled charger to supply 30 A to a 36 VDC battery pack from a 50 VAC source if the series resistance is 0.2 Ω.

Solution

A rearrangement of Equation 3.45 results in the transcendental equation

$$\cos\alpha + \alpha\gamma = \frac{2\pi R I_b}{V_m} + \cos\beta + \beta\gamma. \tag{3.46}$$

ON THE CD

Equation 3.46 is solved with the *alphasolv* function in folder Chapter 3\Toolbox on the CD-ROM. The function is accessed by program Example3_8.m in folder Chapter 3\Examples and Listings. The computed delay angle is 65°.

Conclusion

A thyristor delay angle in the range specified by Equation 3.42 limits the average battery current.

EXERCISE 3.8

Determine the maximum possible charging current of the battery in Example 3.8.

Answer

$I_b = 37.5$ A.

Power and Efficiency

The series resistance of the HW charger is not considered part of the load: it represents the combined resistance of the supply transformer secondary winding and the internal resistance of the battery. The series resistance thus contributes to the circuit power losses. The efficiency is the ratio of the absorbed battery power to the sum of the battery power and power losses:

$$\eta = \frac{P_b}{P_b + P_{SCR} + P_R}. \tag{3.47}$$

The next example illustrates the efficiency computation for the HW phase-controlled charger.

EXAMPLE 3.9

Determine the efficiency of the charger in Example 3.8 if $V_{ak} = 1.2$ V.

Solution

The power absorbed by the battery is $P_b = (36)(30) = 1080$ W. The power dissipated in the thyristor is $P_{SCR} = (1.2)(30) = 36$ W. To determine the power dissipated in the resistor, the RMS value of the battery current is required. The code in Listing 3.4 computes the RMS value of the battery current and the power dissipation in the series resistance.

LISTING 3.4 Code to Compute Efficiency of the HW Phase-Controlled Charger

```
close all, clear all, clc
tpi = 2*pi;
Vs = 50;
Vm = Vs*sqrt(2);
Vb = 36;
Vak = 1.2;
gma = Vb/Vm;
R = 0.2;
alpha = 65;
```

```
a = pi*alpha/180;
B = pi - asin(gma);
th = linspace(a, B, 1024);
Ith = Vm/R*(sin(th) - gma);
Irms = sqrt(1/tpi*trapz(th, Ith.^2))
Ib = 1/tpi*trapz(th, Ith);
Pr = Irms^2 * R
Pb = Vb*Ib
Pscr = Vak*Ib
eta = Pb/(Pb + Pscr + Pr)
```

The RMS value computed by the code is 66.5 A, the power dissipation in the series resistance is 884 W, and the efficiency is 54%.

Conclusion

The power dissipation in the series resistance appears unreasonably large and, in fact, it is. Although the power dissipation is shared by the battery and the transformer, a practical circuit has series inductance that limits the RMS value of the battery current.

EXERCISE 3.9

Determine the efficiency of the charger in Example 3.9 if the battery is charged at the maximum possible current.

Answer

$\eta = 55.6\%$.

FW Phase-Controlled Charger with Series Resistance

The FW phase-controlled charger with series resistance is illustrated in Figure 3.14. The instantaneous battery current and the battery voltage are plotted in Figure 3.15.

As indicated in Figure 3.15, the battery current of the FW circuit has a period of π. The average battery current of the FW charger is thus twice that of the HW charger:

$$I_b = \frac{V_m}{\pi R}\Big[\cos\alpha - \cos\beta - \gamma(\beta - \alpha)\Big]. \tag{3.48}$$

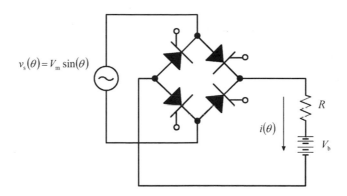

FIGURE 3.14 FW phase-controlled charger with series resistance.

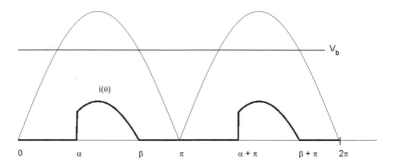

FIGURE 3.15 Instantaneous battery current of the FW charger with series resistance.

The minimum possible delay angle for the FW circuit is the same as for the HW circuit.

EXAMPLE 3.10

Repeat Example 3.8 for a full-wave phase-controlled rectifier.

Solution

With a period of π, Equation 3.46 becomes

$$\cos\alpha + \alpha\gamma = \frac{\pi R I_b}{V_m} + \cos\beta + \beta\gamma. \tag{3.49}$$

The same *alphasolv* function used for the HW circuit is used for the FW circuit with an average current specification of half the required value. The computation is performed by program file Example3_10.m in folder Chapter 3\Examples and Listings on the CD-ROM. The computed delay angle is 97.8°.

Conclusion

A larger delay angle (smaller conduction angle) is required for the FW circuit because the average current is twice that of the HW charger.

EXERCISE 3.10

Determine the maximum charging current for the FW circuit in Example 3.10.

Answer

$I_b = 75$ A.

Power and Efficiency

The efficiency expression for the HW circuit is readily modified for the FW circuit since there are four thyristors in the FW charger:

$$\eta = \frac{P_b}{P_b + 4P_{SCR} + P_R}. \tag{3.50}$$

The power dissipation in the thyristors is computed with half the average battery current because only two thyristors conduct current for each half-cycle of the source voltage. The next example illustrates the efficiency computation for the FW charger with series resistance.

EXAMPLE 3.11

Determine the efficiency of the FW charger in Example 3.10 if $V_{ak} = 1.2$ V.

Solution

Since the average battery current is the same as in Example 3.8, the power absorbed by the battery is 1080 W. The power dissipated in one thyristor is $P_{SCR} = (1.2)(15) = 18$ W. The RMS value of the battery current is readily computed with the code in Listing 3.2, with a change of the delay angle to 97.8° and the period changed from 2π to π. The RMS value that results is 62.2 A, and the power dissipated in the series resistance is 775 W. The efficiency is 56%. The program that performs the computations is available on the CD-ROM as Example3_11.m in folder Chapter 3\Examples and Listings.

Conclusion

Although there is more power dissipation in the thyristors, the decreased RMS current results in a lower series resistance dissipation and slightly better efficiency.

EXERCISE 3.11

Determine the efficiency of the charger in Example 3.11 if the battery is charged at the maximum possible current.

Answer

$\eta = 54.6\%$.

HW Phase-Controlled Charger with Series R-L Impedance

The previous phase-controlled charger circuits were modeled without reactance. The model is accurate for relatively low charging currents and high series resistances. In higher current applications, the energy stored in the series inductance becomes significant and results in longer thyristor conduction times. The HW phase-controlled charger with series R-L impedance is shown in Figure 3.16.

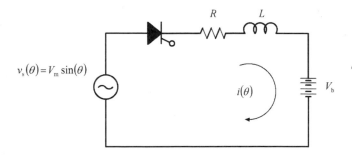

FIGURE 3.16 HW phase-controlled charger with series R-L impedance.

Application of Kirchhoff's voltage law around the circuit in Figure 3.16 yields the same differential equation as that of the diode charger with series R-L impedance presented in Chapter 2. The general solution to the differential equation is the same as well:

$$i(\theta) = A_0 e^{-\theta/q} + \frac{V_m}{R}\cos\phi\sin(\theta-\phi) - \frac{V_b}{R}. \tag{3.51}$$

The initial condition for the controlled-rectifier circuit is

$$i(\alpha) = 0. \tag{3.52}$$

As with the previous charger circuits, the minimum delay angle is expressed by Equation 3.43. Application of the initial condition to Equation 3.51 results in

$$A_0 e^{-\alpha/q} + \frac{V_m}{R}\cos\phi\sin(\alpha-\phi) - \frac{V_b}{R} = 0. \tag{3.53}$$

The solution of Equation 3.53 for the unknown constant A_0 is

$$A_0 = \frac{V_m}{R}e^{\alpha/q}\Big[\, \gamma - \cos\phi\sin(\alpha-\phi)\Big]. \tag{3.54}$$

Substitution of Equation 3.54 into Equation 3.51 yields the expression for the instantaneous current of the HW phase-controlled charger with series R-L impedance:

$$i(\theta) = \frac{V_m}{R}\Big\{\Big[\,\gamma - \cos\phi\sin(\alpha-\phi)\Big]e^{-(\theta-\alpha)/q} + \cos\phi\sin(\theta-\phi) - \gamma\Big\},\ \alpha \le \theta \le \beta. \tag{3.55}$$

The instantaneous battery current is plotted in Figure 3.17.

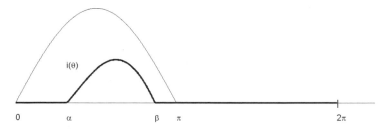

FIGURE 3.17 Battery current of the HW phase-controlled charger with series R-L impedance.

The extinction angle β is determined from Equation 3.55 when the battery current is zero:

$$\Big[\cos\phi\sin(\beta-\phi) - \gamma\Big]e^{\beta/q} = \Big[\cos\phi\sin(\alpha-\phi) - \gamma\Big]e^{\alpha/q}. \tag{3.56}$$

The next example illustrates the average battery computation by numerical integration.

EXAMPLE 3.12

Determine the average battery current supplied by a half-wave phase-controlled charger to a 36 VDC battery pack from a 50 VAC source if the series impedance is $z = 0.2 + j0.2 \, \Omega$ and the delay angle is 65°.

Solution

A numerical solution for the average battery current is more tractable because of the lengthy expression for the instantaneous current. The solution is provided by the MATLAB code in Listing 3.5 and is available on the CD-ROM as Example3_12.m in folder Chapter 3\Examples and Listings.

ON THE CD

LISTING 3.5   Code to Compute Average Battery Current of the HW Charger with R-L Impedance

```
close all, clear all, clc
Vs = 50;
f = 60;
Vm = Vs*sqrt(2);
Vb = 36;
R = 0.2;
X = 0.2;
q = X/R;
phi = atan(q);
gma = Vb/Vm;
alpha = 65;
a = pi*alpha/180;
B = beta_RL(a, q, gma);
th = linspace(a, B, 1024);
cph = cos(phi);
A0 = Vm/R*exp(a/q)*(gma - cph*sin(a - phi));
Ith = A0*exp(-th/q) + Vm/R*cph*sin(th - phi) - Vb/R;
Ib = 1/(2*pi)*trapz(th, Ith)
```

Conclusion

The computed average battery current is 21.1 A. The code in Listing 3.5 uses the *beta_RL* function to compute the extinction angle. The function is available in the Toolbox subfolder of Chapter 3 on the CD-ROM.

ON THE CD

   A comparison of Examples 3.8 and 3.12 illustrates the current-limiting effect of the inductive reactance; the battery current is reduced from 30 to 21 A.

EXERCISE 3.12

Determine the maximum charging current for the battery in Example 3.11.

Answer

$I_b = 27.8$ A.

Efficiency

The efficiency of the HW charger with series R-L impedance is computed with Equation 3.47 since there is no dissipation in the inductive reactance. The power and efficiency computations are readily performed by a modification of the code in Listing 3.4.

FW Phase-Controlled Charger with Series R-L Impedance

The final circuit presented in this chapter is the FW charger with series R-L impedance. The schematic is shown in Figure 3.18.

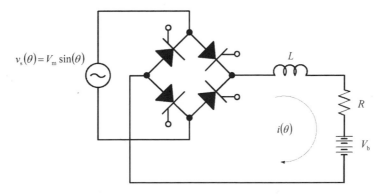

FIGURE 3.18   FW charger with series R-L impedance.

The instantaneous current of the circuit in Figure 3.18 is expressed by Equation 3.55; however, the period of the waveform is π rather than 2π.

As the earlier circuit without the battery, the FW charger with series R-L impedance has discontinuous and continuous current modes of operation.

Discontinuous Battery Current

Discontinuous operation is nearly identical to that of the HW charger; the exception is the period of the current waveform, which is π for the FW circuit. The extinction angle is computed with Equation 3.56 and is less than $\alpha + \pi$ when the battery current is discontinuous. The instantaneous battery current is plotted in Figure 3.19.

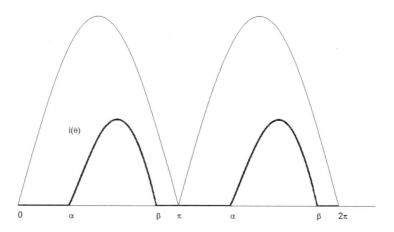

FIGURE 3.19 Discontinuous battery current of the FW phase-controlled charger with R-L impedance.

As with the HW charger, a numerical solution for the average battery current is more tractable than a closed-form solution. The computation is readily performed by the MATLAB code in Listing 3.5, with the period changed to π.

Continuous Battery Current

With sufficient inductance and smaller γ-ratios, the battery current becomes continuous. The minimum inductance for continuous battery current corresponds to a value of $\alpha + \pi$ for the extinction angle. Substitution of $\beta = \pi + \alpha$ into Equation 3.56 yields the transcendental equation

$$\left[\cos\phi\sin\phi - \gamma\right]e^{\pi/q} = \left[\cos\phi\sin(\alpha - \phi) - \gamma\right]e^{\alpha/q}. \tag{3.57}$$

The unknown variable in Equation 3.57 is the quality factor q, which is a function of the inductance. Although Equation 3.57 is transcendental, the numerical solution is more tractable if the expression is rearranged into a known lefthand side (LHS) and an unknown righthand side (RHS):

$$\pi - \alpha = q \ln \left[\frac{2\cos\phi\sin(\alpha - \phi) - 2\gamma}{\sin 2\phi - 2\gamma} \right].$$ (3.58)

The next example illustrates a numerical solution of Equation 3.58.

EXAMPLE 3.13

Write a MATLAB function to compute the quality factor of the continuous-current FW phase-controlled charger with input arguments α and γ.

Solution

ON THE CD

The solution is provided in Listing 3.6 and is available as *qsolv.m* in the Toolbox subfolder of Chapter 3 on the CD-ROM.

LISTING 3.6   Function to Compute q from α and γ

```
function q = qsolv(a, gma)
amin = asin(gma)
if a < amin,
    disp('alpha must be greater than arcsin(Vb/Vm)')
    break
end
RHS = 1;
q = 120*pi;
inc = q/2;
phi = atan(q);
while(abs(RHS) > 1E-6),
    RHS = q*log((2*cos(phi)*sin(a - phi) - 2*gma)/...
        (sin(2*phi) - 2*gma)) - (pi - a);
    if RHS > 0,
        q = q - inc;
    else
        inc = inc/2;
        q = q + inc;
    end
end
```

Conclusion

The function checks the delay angle to ensure that it is greater than the minimum value; otherwise, an erroneous value of quality factor results. The first value of q attempted for the solution is one that corresponds to a 60 Hz source and an inductive time constant of 1 second. The value is decreased until *RHS* changes sign.

After the sign change, the increment is decreased and added to q; the smaller increment is added until another sign change occurs. The process continues until smaller increments result in a value of q within the tolerance specified by the *while* statement. The inductance is determined from the computed value of q with the relationship

$$L = \frac{qR}{\omega}.$$ (3.59)

EXERCISE 3.13

Determine the minimum inductance required for continuous battery current operation for the FW charger with $V_m = 170$, $V_b = 120$, $R = 0.8$, $f = 60$ Hz, and $\alpha = \pi/3$.

Answer

$L = 0.78$ H.

Continuous Battery Current

When the battery current is continuous, Equation 3.51 is subject to the same boundary condition expressed in Equation 3.31. By the principle of superposition, the instantaneous continuous battery current is

$$i(\theta) = \frac{V_m}{R}\cos\phi\left[\sin(\theta - \phi) - \sin(\alpha - \phi)\operatorname{csch}\left(\frac{\pi}{2q}\right)e^{-(\theta - \chi)/q}\right] - \frac{V_b}{R}.$$ (3.60)

The continuous battery current of the FW phase-controlled charger with R-L impedance has the same waveform shape as illustrated in Figure 3.11.

EXAMPLE 3.14

Specify a transformer for a full-wave phase-controlled charger to supply current to a 144 VDC battery pack from a 120 VAC, 60 Hz voltage source. The combined series impedance of the transformer secondary winding and the battery is $Z = 0.2 + j0.06\ \Omega$.

Solution

The transformer is specified by the volt-amperes of the load. The maximum volt-amperes occurs at the minimum delay angle expressed by Equation 3.43 which, for this example, is approximately 60°.

To proceed with the solution, the mode of operation, continuous or discontinuous battery current, must be determined. The quality factor of the circuit is $q = 0.06/0.1 = 0.6$. With the *qsolv* function, the quality factor for continuous current is computed as $q = 447$. Since the actual q of the circuit is less, the charger operates in the discontinuous current mode. The code shown in Listing 3.7 computes the RMS current drawn by the charger followed by the calculation of apparent power of the load. The program is available as Example3_14.m in subfolder Examples and Listings of Chapter 3 on the CD-ROM.

ON THE CD

LISTING 3.7 Code to Compute the Apparent Load Power of the FW Phase-Controlled Charger

```
close all, clear all, clc
Vs = 120;
f = 60;
Vm = Vs*sqrt(2);
Vb = 144;
R = 0.2;
X = 0.06;
q = X/R;
phi = atan(q);
gma = Vb/Vm;
alpha = 60;
a = alpha*pi/180;
B = beta_RL(a, q, gma);
th = linspace(a, B, 1024);
cphi = cos(phi);
Ith = Vm/R*((gma - cphi*sin(a - phi))*exp(-(th - a)/q)...
       + cphi*sin(th - phi) - gma);
Is = sqrt(1/pi*trapz(th, Ith.^2))
S = Vs*Is
```

Conclusion

The computed result of the code in Listing 3.7 is $S = 5.6$ kVA. The large quality factor required for continuous current operation requires a very large series inductance that is not likely to be encountered in practice. Hence, discontinuous battery current is the typical mode of operation for the FW phase-controlled charger with R-L impedance.

EXERCISE 3.14

Determine the average battery current of the charger in Example 3.14.

Answer

$I_b = 26.3$ A.

CHAPTER SUMMARY

The advantage of the phase-controlled rectifier over the diode rectifier is the variable average output current of the controlled-rectifier circuit. The variable output current provides control of heating elements (resistive and *R-L* loads) and control of battery-charging currents. Although the diode charger has an element of self-regulation of current, the variable current of the controlled rectifier provides more precise control. The design of the supply transformer for the diode charger is critical because its properties influence the self-regulating mechanism. A commercially available transformer may be used in the controlled charger since the control mechanism is the thyristor delay angle.

The theory developed in this chapter is also applicable to the control of DC motors with a back-electromotive force (EMF) that is proportional to motor speed. In the charger circuit models the battery voltage is simply replaced by the back EMF expression of the DC motor. In these control circuits the adjustable delay angle controls the speed of the DC motor.

ON THE CD

THE MATLAB TOOLBOX

The functions listed below are located in the Toolbox subfolder of Chapter 3 on the CD-ROM.

HWCR

Function *hwcr.m* computes the average output current of the HW phase-controlled rectifier with resistive load from the RMS source voltage, the load resistance, and the thyristor delay angle. Also computed are the apparent, real, reactive, and distortive powers and the displacement, distortion, and power factors.

Syntax

```
[Io, S, P, Q, D, DPF, DF, pf] = hwcr(Vs, R, alpha)
```

FWCR

Function *fwcr.m* computes the average output current of the FW phase-controlled rectifier with resistive load from the RMS source voltage, the load resistance, and the thyristor delay angle. Also computed are the apparent, real, reactive, and distortive powers and the displacement, distortion, and power factors.

Syntax

```
[Io, S, P, Q, D, DPF, DF, pf] = fwcr(Vs, R, alpha)
```

HWCRL

Function *hwcrl.m* computes the average output current of the HW phase-controlled rectifier with *R-L* load from the RMS source voltage, the load resistance, the inductive reactance, and the thyristor delay angle. Also computed are the apparent, real, reactive, and distortive powers and the displacement, distortion, and power factors.

Syntax

```
[Io, S, P, Q, D, DPF, DF, pf] = hwcrl(Vs, R, X, alpha)
```

FWCRL

Function *fwcrl.m* computes the average output current of the FW phase-controlled rectifier with *R-L* load from the RMS source voltage, the load resistance, the inductive reactance, and the thyristor delay angle. Also computed are the apparent, real, reactive, and distortive powers and the displacement, distortion, and power factors.

Syntax

```
[Io, S, P, Q, D, DPF, DF, pf] = fwcrl(Vs, R, X, alpha)
```

HWCRV

Function *hwcrv.m* computes the average output current of the HW phase-controlled charger from the RMS source voltage, series resistance, battery voltage, and thyristor delay angle. Also computed are the apparent, real, reactive, and distortive powers along with the displacement, distortion, and power factors.

Syntax

```
[Io, S, P, Q, D, DPF, DF, pf] = hwrv(Vs, R, Vb, alpha)
```

FWCRV

Function *fwcrv.m* computes the average output current of the FW phase-controlled charger from the RMS source voltage, series resistance, battery voltage, and thyristor delay angle. Also computed are the apparent, real, reactive, and distortive powers along with the displacement, distortion, and power factors.

Syntax

```
[Io, S, P, Q, D, DPF, DF, pf] = fwcrv(Vs, R, Vb, alpha)
```

HWCRLV

Function *hwcrlv.m* computes the average output current of the HW charger with series *R-L* impedance from the RMS source voltage, series resistance, inductive reactance, battery voltage, and thyristor delay angle. Also computed are the apparent, real, reactive, and distortive powers along with the displacement, distortion, and power factors.

Syntax

```
[Io, S, P, Q, D, DPF, DF, pf] = hwcrlv(Vs, R, X, Vb, alpha)
```

FWCRLV

Function *fwcrlv.m* computes the average output current of the FW charger with series *R-L* impedance from the RMS source voltage, series resistance, inductive reactance, battery voltage, and thyristor delay angle. Also computed are the apparent, real, reactive, and distortive powers along with the displacement, distortion, and power factors.

Syntax

```
[Io, S, P, Q, D, DPF, DF, pf] = fwcrlv(Vs, R, X, Vb, alpha)
```

PROBLEMS

Problem 1

Compute the delay angle required for an HW phase-controlled rectifier to deliver an average current of 10 A to a 5 Ω load from a 120 VAC source.

Problem 2

Compute the delay angle required for an FW phase-controlled rectifier to deliver an average current of 10 A to a 5 Ω load from a 120 VAC source.

Problem 3

Compute the delay angle required for an HW phase-controlled rectifier to deliver an average current of 10 A to a 5 + j2 Ω load from a 120 VAC, 60 Hz source.

Problem 4

Compute the delay angle required for an FW phase-controlled rectifier to deliver an average current of 10 A to a 5 + j2 Ω load from a 120 VAC, 60 Hz source.

Problem 5

Compute the delay angle required for an HW phase-controlled rectifier to deliver an average current of 10 A to a 36 V battery pack with 0.2 Ω series resistance from a 120 VAC source.

Problem 6

Compute the delay angle required for an FW phase-controlled rectifier to deliver an average current of 10 A to a 36 V battery pack with 0.2 Ω series resistance from a 120 VAC source.

Problem 7

Compute the delay angle required for an HW phase-controlled rectifier to deliver an average current of 10 A to a 36 V battery pack with 0.2 + j0.3 Ω series impedance from a 120 VAC, 60 Hz source.

Problem 8

Compute the delay angle required for an FW phase-controlled rectifier to deliver an average current of 10 A to a 36 V battery pack with $0.2 + j0.3$ Ω series impedance from a 120 VAC, 60 Hz source.

Problem 9

Compute the reactive power and power factor of the circuit in Problem 2. Compute the value of shunt capacitance to neutralize the reactive power along with the improved power factor.

Problem 10

Compute the reactive power and power factor of the circuit in Problem 4. Compute the value of shunt capacitance to neutralize the reactive power along with the improved power factor.

Problem 11

Compute the reactive power and power factor of the circuit in Problem 6. Compute the value of shunt capacitance to neutralize the reactive power along with the improved power factor.

Problem 12

Compute the reactive power and power factor of the circuit in Problem 8. Compute the value of shunt capacitance to neutralize the reactive power along with the improved power factor.

Part III

DC-DC Converter Technology

Chapter 2 presented circuits that convert AC to DC. The rectifier circuit provides an average voltage to a type of load that requires a direct current. After rectification, it is often necessary to provide more than one DC voltage to various devices or equipment. For example, a system might require 12 VDC for a motor, 5 VDC for logic circuitry, and possibly −12 VDC for a communications channel. One solution to this design problem is to provide three separate rectifier circuits, each with its own transformer with the proper turns ratio to provide the required DC level. This solution is applicable when separate *isolated* outputs are required. Often, however, the output potentials are with respect to one another and, in such a case, a multitapped transformer applies. In either case, the size, weight, and cost of the transformers may not provide an optimum design. Furthermore, the issue of *regulation* must be addressed.

The circuit technology that meets the design challenges of size, weight, and cost are DC-DC converters. These circuits take a DC voltage input and convert it to a higher or lower DC voltage output, or even a DC voltage output of opposite polarity than the input. Furthermore, these circuits provide regulation, as they keep the output voltage constant despite changes in the input voltage and the output current.

4

Introduction to DC-DC Converters: the Buck Converter

In This Chapter

- Introduction
- Efficiency and Regulation
- The Buck Converter Circuit
- DC-DC Converter Analysis
- The Continuous Conduction Mode
- The Discontinuous Conduction Mode
- Practical Buck Converters
- Chapter Summary
- The MATLAB Toolbox
- Problems

INTRODUCTION

The DC-DC converter is an electrical circuit that transfers energy from a DC voltage source to a load. The energy is first transferred via electronic switches to energy storage devices and then subsequently switched from storage into the load. The switches are transistors and diodes; the storage devices are inductors and capacitors. This process of energy transfer results in an output voltage that is related to the input voltage by the duty ratios of the switches.

It is insightful and worthwhile to investigate why DC-DC converters are necessary before their detailed presentation and analysis. In addition to the constraints of size, weight, and cost, DC-DC converter technology also addresses the issues of efficiency and regulation.

EFFICIENCY AND REGULATION

Consider a power supply design problem in which 5 VDC at 4 A must be provided from a 12 VDC supply. One solution is to use a resistance to drop the 12 V down to 5 V. The required resistance is $(12 - 5)/4 = 1.75\ \Omega$. The load resistance is $5/4 = 1.25\ \Omega$. The circuit is shown in Figure 4.1.

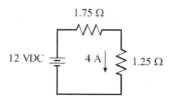

FIGURE 4.1 Circuit to provide 5 VDC at 4 A from a 12 VDC source.

Two major issues are associated with the circuit in Figure 4.1. The first is efficiency. The power required by the load is $(5)(4) = 20$ W; however, the power that must be supplied by the source is $(12)(4) = 48$ W, which results in an efficiency of only 20/48—approximately 42%. The low efficiency is due to the power lost in the dropping resistor: $(4^2)(1.75) = 28$ W (or $48 - 20 = 28$ W). More energy is lost in the delivery of the power than is absorbed by the load.

The second problem with the circuit in Figure 4.1 is the lack of regulation. It is rare that a primary source is absolutely constant. For example, the voltage of a 12 V automotive starting battery can vary as much as ± 2 V, depending on whether the battery is being charged (14 V) or if the car is started while the head and tail lights are on (10 V). If the source in the circuit in Figure 4.1 drops to 10 V, the output voltage would drop to $(10)(1.25/3) \approx 4.2$ V. If the source is increased to 14 V, the output increases to $(14)(1.25/3) \approx 5.8$ V. These voltage fluctuations could cause severe functional problems for the load, especially if the load is a circuit based on 5 V logic technology.

To overcome the efficiency and regulation problems, a control scheme must be introduced whereby the output voltage is measured and the supplied energy is varied in accordance with the requirements of the load. If the source voltage increases, the controller must decrease the source current to prevent an increase in load voltage. A decreased source voltage requires an increase of output current to keep the output voltage constant.

Linear Regulators

Linear regulation makes use of an amplifier and a transistor to control the current supplied to the load. Figure 4.2 illustrates a linear regulator circuit in which a PNP transistor replaces the 1.75 Ω resistor in the circuit in Figure 4.1.

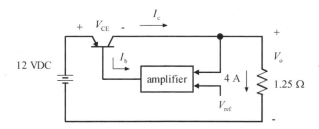

FIGURE 4.2   Linear regulator circuit.

The amplifier in the circuit draws base current from the PNP transistor. The gain of the amplifier is set such that the base current is decreased when the error voltage increases and vice versa. The error voltage is the difference between the output voltage and the reference voltage, $V_o - V_{ref}$. If the error voltage increases, which corresponds to an increase in V_o, the base current is reduced. The decreased base current causes a proportionate decrease in the collector current, which is also the load current. If $V_o - V_{ref}$ decreases, which corresponds to a decrease in V_o, the base current is increased in order to supply more load current and maintain the output voltage constant.

The linear regulator effectively solves the regulation problem; however, the efficiency is still low because the transistor must operate in the active region and drop the source voltage down to the load voltage. The transistor essentially dissipates the same power as the dropping resistor in the unregulated circuit. To operate efficiently, the transistor must be operated in the saturation region where the collector-to-emitter voltage, V_{CE}, is relatively small.

Pulse-Width-Modulation

A control method by which the transistor operates in either saturation or cutoff is called pulse-width modulation (PWM). The details of a PWM controller are considered later; it is beneficial to first gain an understanding of how a circuit functions under PWM control.

When controlled by PWM, the transistor is saturated for a length of time known as the *on-time*; it is then driven to the cutoff region for a time called the *off-time*. During the on-time V_{CE} is very small in comparison to the other voltages in the circuit and is typically approximated as zero to simplify the circuit analysis. During the off-time the transistor is cut off, which effectively results in an open circuit between the collector and the emitter. The transistor thus operates as a switch that is closed during the on-time and open during the off-time. When the switch is closed, the full source voltage appears across the load; when the switch is opened, the load voltage is zero. Figure 4.3 illustrates the states of the PWM-controlled circuit along with the instantaneous output voltage.

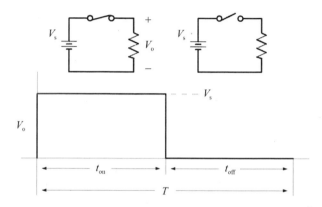

FIGURE 4.3   States of the PWM-controlled circuit and the instantaneous output voltage.

As illustrated in the figure, the sum of the on- and off-times is the period T of the switch. The switching action repeats every T seconds. The reciprocal of the period is the switching frequency, that is, the rate at which the transistor is switched:

$$f = \frac{1}{T}.$$

(4.1)

The voltage across the load is obviously not constant; however, it does have a DC component. The average value of the output voltage is determined from

$$V_o = \frac{1}{T} \int_0^{t_{on}} V_s \, dt.$$

(4.2)

Evaluation of the integral in Equation 4.2 leads to

$$V_o = \frac{t_{on}}{T} V_s.$$

(4.3)

Equation 4.3 reveals that the average output voltage is proportional to the on-time. The longer the switch stays closed, the larger the DC component of the output voltage. The ratio of the on-time to the switching period is called the duty ratio:

$$D = \frac{t_{on}}{T}. \tag{4.4}$$

In terms of Equation 4.4, the average output voltage is expressed as

$$V_o = DV_s. \tag{4.5}$$

As the duty ratio varies between zero and one, the average output voltage varies between zero and the source voltage.

The PWM method is also effective as a linear controller of the average power delivered to the load. The RMS value of the output voltage is determined from

$$V_o = \sqrt{\frac{1}{T} \int_0^{t_{on}} V_s^2 dt}. \tag{4.6}$$

Substitution of Equation 4.5 into Equation 4.6 and evaluation of the integral results in an RMS value of

$$V_{RMS} = V_s \sqrt{D}. \tag{4.7}$$

The average power absorbed in the PWM-controlled load is thus

$$P = D\frac{V_s^2}{R}. \tag{4.8}$$

The PWM-controlled circuit offers the advantage of linear control over the load power. It is quite effective in the control of the intensity of incandescent lighting and light-emitting diode (LED) displays, as well as the energy applied to heating elements. These types of devices suffer no functional problems when the current is periodically interrupted. However, in the case of the logic circuit described earlier, an interruption of supply current results in a complete loss of functionality. In order to provide an effectively constant output, the circuit in Figure 4.3 must be modified with the addition of energy-storage components that will sustain the output voltage during the off-time of the power switch. A circuit that provides a constant output voltage that is lower in magnitude than the input voltage is called a buck converter, or step-down converter.

THE BUCK CONVERTER CIRCUIT

The buck converter, shown in Figure 4.4, has a capacitor and an inductor along with two *complementary* switches; when one switch is closed, the other is open and vice versa.

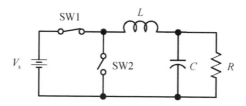

FIGURE 4.4   Buck regulator circuit.

The switches are alternately opened and closed at the rate of the PWM switching frequency. The output that results is a regulated voltage of smaller magnitude than the source voltage.

When switch SW1 is closed (SW2 open), the source delivers energy to the R-L-C network. The inductor and capacitor store energy during this time interval, and the resistive load absorbs energy. When switch SW2 is closed (SW1 is open), the source is disconnected from the network and the inductor terminal that was previously connected to the source is now short-circuited to ground. During this second interval a portion of the energy stored in L and C is absorbed by the load. One switching period is completed when SW2 opens and SW1 closes again. The on-time for the circuit is defined with respect to SW1. Hence, SW2 is on during the off-time of SW1. The duty ratio of SW1 is therefore D, and the duty ratio of SW2 is $1 - D$. The switching period of the circuit is illustrated in Figure 4.5, in which the equivalent circuits are shown in accordance with the switch positions.

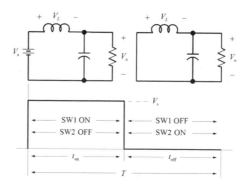

FIGURE 4.5   Buck converter equivalent circuits.

Since the purpose of the circuit is to produce a regulated output, a certain amount of inductance and capacitance is required to sustain the energy flow to the resistive load and maintain the output voltage during the off-time of SW1. These values of L and C are determined from an analysis of the circuit.

DC-DC CONVERTER ANALYSIS

There are various methods of analysis for DC-DC converters. One approach is to derive the differential equations that describe the inductor current and capacitor voltage and then solve them in accordance with the boundary conditions of the switching period. This exacting method, although accurate, produces a cumbersome set of equations that require extensive computation and is best reserved for simulation purposes.

Another approach is based on the assumption that the output voltage is essentially constant throughout the switching period. This method greatly simplifies the mathematics; however, the method neglects the dynamic behavior of the circuit. Although the simplified approach is less accurate, it nevertheless yields an excellent body of theory that is widely applicable in DC-DC converter design.

The analysis begins with a formulation of the instantaneous inductor current. A development of the duty ratio follows, where an expression is obtained in terms of the input and output voltages. Finally, the assumption of absolutely constant output voltage is relaxed and a formula for capacitance is derived from a specified amount of allowable output voltage variation.

Inductor Current

The current through a linear inductor is expressed by the integral equation

$$i_L(t) = \frac{1}{L} \int_{t_0}^{t} v_L(\tau)\,d\tau + I_{\min}, \tag{4.9}$$

in which $v_L(\tau)$ is the voltage across the inductor and $I_{\min}$ is the value of inductor current at $t = t_0$. The voltage across the inductor in the buck converter is determined by Kirchhoff's voltage law (KVL) applied to the equivalent circuits in Figure 4.5. In the time interval between 0 and t_{on}, the inductor voltage is

$$V_{L,1} = V_s - V_o, \tag{4.10}$$

and from time t_{on} to T, the voltage is

$$V_{L,2} = -V_o. \tag{4.11}$$

The instantaneous inductor voltage is illustrated in Figure 4.6.

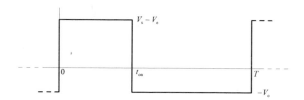

FIGURE 4.6 Instantaneous inductor voltage.

Substitution of Equations 4.10 and 4.11 into Equation 4.9 yields the inductor current expressions

$$i_{L,1}(t) = \frac{1}{L} \int_0^t (V_s - V_o) d\tau + I_{min}, \, 0 \le t \le t_{on} \quad (4.12)$$

and

$$i_{L,2}(t) = \frac{1}{L} \int_{t_{on}}^t (-V_o) d\tau + I_{max}, \, t_{on} \le t \le T. \quad (4.13)$$

The solutions to the integral expressions in Equations 4.12 and 4.13 are

$$i_{L,1}(t) = \frac{V_s - V_o}{L} t + I_{min} \quad (4.14)$$

and

$$i_{L,2}(t) = \frac{V_o}{L}(T - t) + I_{min}. \quad (4.15)$$

The instantaneous inductor current is illustrated in Figure 4.7.

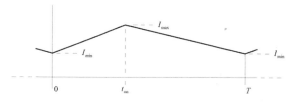

FIGURE 4.7 Instantaneous inductor current.

Inductance

There are further implications that arise from the assumption that the capacitance is large enough to maintain the output voltage constant. The expression "large enough" essentially means infinite. As the value of capacitance approaches infinity, the capacitive reactance, $-\frac{1}{\omega C}$, approaches zero. The capacitor thus appears as a short circuit to all the harmonic content of the inductor current. The average value of inductor current, however, is blocked by the capacitor but flows freely into the load resistance. This diversion of the alternating components of the inductor current into the capacitor and the average value into the resistor is illustrated in Figure 4.8.

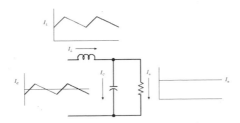

FIGURE 4.8   Flow of inductor current into the capacitor and load resistance.

Since only the average value of inductor current flows into the load, it is the DC component of the inductor current that defines the load current. The average value is obtained from the rectangular and triangular areas between the inductor current waveform and the horizontal time-axis. The waveform geometry is illustrated in Figure 4.9.

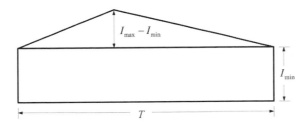

FIGURE 4.9   Geometry of the inductor current waveform.

The total area under the waveform is

$$A = TI_{\min} + \frac{1}{2}T\left(I_{\max} - I_{\min}\right). \tag{4.16}$$

Division of the area expressed in Equation 4.16 by the period T results in the average value

$$I_{avg} = \frac{1}{2}\left(I_{max} - I_{min}\right) + I_{min}. \tag{4.17}$$

The peak current is obtained from Equation 4.14 with $t = t_{on}$.

$$I_{max} = \frac{V_s - V_o}{L}t_{on} + I_{min}. \tag{4.18}$$

Substitution of Equation 4.18 into Equation 4.17 results in an average value of

$$I_{avg} = \frac{V_s - V_o}{2L}t_{on} + I_{min}. \tag{4.19}$$

With Equations 4.1 and 4.4, the on-time is expressed as

$$t_{on} = \frac{D}{f}. \tag{4.20}$$

Substitution of Equation 4.20 into Equation 4.19 yields the expression for the average inductor current in terms of the duty ratio and the PWM switching frequency:

$$I_{avg} = \frac{D\left(V_s - V_o\right)}{2Lf} + I_{min}. \tag{4.21}$$

As stated earlier, it is only the average value of $i_L(t)$ that flows into the load resistance. Consequently, the average inductor current is also the output current:

$$I_{avg} = \frac{V_o}{R}. \tag{4.22}$$

Substitution of Equation 4.22 into 4.21 results in the load current expression

$$\frac{V_o}{R} = \frac{D\left(V_s - V_o\right)}{2Lf} + I_{min}. \tag{4.23}$$

The solution of Equation 4.23 for the inductance is

$$L = \frac{D(V_s - V_o)R}{2f(V_o - I_{min}R)}. \tag{4.24}$$

Equation 4.24 is the desired result of the analysis: it expresses a value of inductance calculated from desired voltage and current levels, the duty ratio, the load resistance, and the PWM switching frequency.

The initial inductor current at the beginning of the switching cycle has a significant effect on the behavior of the converter. If $I_{min} > 0$, the inductor current is continuous throughout the switching period and the converter operates in the *continuous conduction mode* (CCM). If the initial current is zero, the inductor current becomes zero before the end of the period; the inductor current is thus discontinuous and the circuit operates in the *discontinuous conduction mode*.

THE CONTINUOUS CONDUCTION MODE

With sufficient inductance, the circuit is made to function in CCM for a desired range of operating conditions. This value of inductance is such that I_{min} is always positive.

CCM Inductance

From Equation 4.23, the expression for the minimum current is

$$I_{min} = \frac{V_o}{R} - \frac{D(V_s - V_o)}{2Lf}. \tag{4.25}$$

For CCM operation, the inequality

$$\frac{V_o}{R} - \frac{D(V_s - V_o)}{2Lf} > 0 \tag{4.26}$$

must be satisfied. The solution of Equation 4.25 for L when $I_{min} = 0$ yields the expression for the minimum inductance required for CCM operation:

$$L_{CCM} = \frac{D(V_s - V_o)R}{2fV_o}. \tag{4.27}$$

The value of CCM inductance computed from Equation 4.27 results in a minimum inductor current of zero. In practice, to ensure CCM operation, a larger inductance is used to provide a margin of error with a guaranteed positive value of I_{min}.

CCM Duty Ratio

The derivation of the duty ratio expression is based on the principle that *the average voltage across an inductor is zero in the periodic steady-state condition*. The voltage across a linear inductor is expressed by

$$V_L = L\frac{di}{dt}. \tag{4.28}$$

Substitution of Equation 4.28 into the definition of the average value yields the expression for the average inductor voltage

$$V_{avg} = \frac{1}{T}\int_0^T \left(L\frac{di}{dt} \right) dt = \frac{1}{T}\int_0^T Ldi. \tag{4.29}$$

Evaluation of the integral in Equation 4.29 leads to

$$V_{avg} = \frac{L}{T}i\Big|_0^T = \frac{L}{T}\Big[i(T)-i(0) \Big]. \tag{4.30}$$

Since the inductor current is periodic, the current at time T is equal to the current at time zero; hence, the average inductor voltage is zero. With zero average voltage, the algebraic sum of the rectangular areas above and below the time-axis in Figure 4.6 must sum to zero:

$$\left(V_s - V_o \right)t_{on} + \left(-V_o \right)\left(T - t_{on} \right) = 0. \tag{4.31}$$

Simplification of Equation 4.31 results in

$$V_s t_{on} = V_o T, \tag{4.32}$$

and, with Equation 4.4, Equation 4.32 simplifies to

$$D = \frac{V_o}{V_s}. \tag{4.33}$$

Equation 4.33 is both the expression that determines the duty ratio *and* the transfer function of the circuit. With the duty ratio expression, the CCM inductance expressed by Equation 4.27 simplifies to

$$L_{\text{CCM}} = \frac{(1-D)R}{2f}. \tag{4.34}$$

Applications of Equation 4.34 are illustrated in the examples and exercises that follow.

EXAMPLE 4.1

A buck regulator is supplied from a 12 VDC source and outputs 5 VDC at 4 A. Determine the CCM inductance if the PWM switching frequency is 20 kHz.

Solution

The duty ratio of the circuit is $D = V_o/V_s = 5/12 \approx 0.42$, or 42%. The load resistance is $R = V_o / I_o = 5/4 = 1.25 \ \Omega$. With Equation 4.34, the CCM inductance is computed as

$$L_{\text{CCM}} = \frac{(1-0.42)(1.25)}{(2)(20\text{E}3)} \approx 18 \ \mu\text{H}.$$

Conclusion

The inductance must be greater than 18 μH to ensure CCM operation. The final choice of inductance used in a design is based on size, weight, and economic factors, as well as other circuit limitations and specifications.

EXERCISE 4.1

Determine L_{CCM} if the load current in Example 4.1 is 2 A. Why is more inductance required for the smaller load current?

Answer

$L_{\text{CCM}} = 36 \ \mu\text{H}$.

EXAMPLE 4.2

The buck converter in Example 4.1 has a load current that varies between 1 and 8 A, while the source voltage may vary between 10 and 14 V. Determine the inductance required to maintain CCM operation under all circuit conditions.

Solution

An inspection of Equation 4.34 reveals that the largest inductance occurs at minimum duty ratio and maximum load resistance. These limits occur at maximum source voltage and minimum load current, respectively. The minimum duty ratio is $D = 5/14 \approx 0.36$. The maximum load resistance is $R = 5/1 = 5\ \Omega$. Hence, the minimum inductance required for CCM operation under all variations is

$$L = \frac{(1-0.36)(5)}{(2)(20\text{E}3)} = 80\ \mu\text{H}.$$

Conclusion

When the source voltage and/or load current vary, as they do in practice, the largest CCM inductance must be used to guarantee CCM operation under all circuit conditions.

EXERCISE 4.2

A buck regulator has an input voltage range of 24 to 32 VDC and an output current range of 5 to 15 A. The output voltage is 12 VDC. Determine the CCM inductance for a switching frequency of 25 kHz.

Answer

$L_{\text{CCM}} = 30\ \mu\text{H}$.

Minimum Inductor Current Specification

Often the inductance is specified by a required minimum inductor current. This specification provides a margin of error inside the CCM region. From Equations 4.23, 4.33, and 4.34, the inductance in terms of $I_{\min}$ is

$$L = \frac{L_{\text{CCM}}}{1 - \dfrac{I_{\min}}{I_{\text{o}}}}, \tag{4.35}$$

in which I_{o} is the load current. Equation 4.35 reveals that if the minimum inductor current is zero, the required inductance is L_{CCM}. As the required minimum current increases, the inductance must also increase; and, as $I_{\min}$ approaches the load current, the inductance approaches infinity.

The Lambda Ratio

A rearrangement of Equation 4.35 produces the ratio:

$$\frac{L}{L_{\text{CCM}}} = \frac{I_o}{I_o - I_{\min}} \equiv \lambda. \tag{4.36}$$

The λ-ratio is the ratio of the actual inductance used in the converter circuit to the minimum required CCM inductance. The ratio also represents the ratio of the load current to the difference between the load current and the minimum inductor current. A λ-value of unity, for example, indicates that $L = L_{\text{CCM}}$ and $I_{\min} = 0$. As the inductance is increased above L_{CCM}, the minimum inductor current also increases and the difference between I_o and $I_{\min}$ decreases. From Equation 4.36, the minimum inductor current expressed in terms of the λ-ratio is

$$I_{\min} = \frac{\lambda - 1}{\lambda} I_o. \tag{4.37}$$

With Equations 4.18, 4.20, 4.33, 4.34, and 4.37, the peak inductor current is also expressed in terms of the λ-ratio as

$$I_{\max} = \frac{\lambda + 1}{\lambda} I_o. \tag{4.38}$$

An additional inductor specification that is sometimes used is the inductor *ripple current*, the difference between the peak and the minimum currents:

$$\Delta I_L = I_{\max} - I_{\min}. \tag{4.39}$$

With Equations 4.37 and 4.38, the ripple current is expressed in terms of the λ-ratio:

$$\Delta I_L = \frac{2}{\lambda} I_o. \tag{4.40}$$

A large value of λ, which corresponds to a large value of inductance, results in a lower inductor ripple current. As λ approaches infinity ($L \to \infty$), the ripple current approaches zero, and both $I_{\max}$ and $I_{\min}$ approach the load current I_o. A low ripple current, which corresponds to a limited peak current, is often required to prevent saturation of the inductor. If the inductor becomes saturated, the inductance value will decrease and DCM operation may result. The examples and exercises that follow illustrate the use of the λ-ratio in the computation of inductance.

EXAMPLE 4.3

Specify the inductance required for a minimum inductor current of 0.5 A for the buck converter in Example 4.2.

Solution

The CCM inductance required for Example 4.2 was found to be 80 μH. From Equation 4.35, the inductance needed for a minimum inductor current of 0.5 A is

$$L = \frac{80}{1 - \dfrac{0.5}{1}} = 160 \ \mu H.$$

Equation 4.36 provides a convenient solution in terms of the λ-ratio:

$$\lambda = \frac{I_o}{I_o - I_{\min}} = \frac{1}{1 - 0.5} = 2.$$

Thus, the required inductance must be twice the CCM inductance.

Conclusion

If the load current varies, the minimum load current must be used to calculate the inductance to guarantee the minimum required inductor current under all specified operating conditions. The λ-ratio provides a convenient method for the computation of inductance based on the minimum or maximum inductor current, or the ripple current.

EXERCISE 4.3

Determine the λ-ratio for the buck converter in Example 4.2 with the specification that the ripple current must not exceed 30% of the maximum load current.

Answer

$\lambda = 7$.

EXAMPLE 4.4

Compute the inductance for the converter in Example 4.2 with the requirement that the inductor current not exceed 10 A to prevent saturation.

Solution

From Equation 4.36, the λ-ratio is

$$\lambda = \frac{I_o}{I_{max} - I_o} = \frac{8}{10 - 8} = 4.$$

Thus, the required inductance is $(4)(80) = 320\ \mu H$.

Conclusion

To limit the peak inductor current, the λ-ratio computation must be based on the maximum expected load current.

EXERCISE 4.4

Determine the largest peak inductor current that will occur for the converter circuit in Example 4.2 with a λ-ratio of 2.5.

Answer

$I_{max} = 11.2$ A.

CCM Capacitance

The inductor current analysis is based on the assumption of infinite capacitance. This requirement must be relaxed to arrive at a practical method to compute a finite value of capacitance from a given specification. That specification is typically the peak-to-peak allowable load-voltage variation known as the *ripple voltage*. The capacitor ripple voltage is determined from an analysis of the capacitor current and the principle of capacitor *charge balance*.

CCM Capacitor Current

As demonstrated earlier, the capacitor current is the sum of all the harmonics of the inductor current, that is, it is the instantaneous inductor current minus the average value:

$$i_{C,1}(t) = \frac{V_s - V_o}{L} t + I_{min} - I_o,\ 0 \le t \le t_{on}. \tag{4.41}$$

$$i_{C,2}(t) = \frac{V_o}{L}\left(T - t\right) + I_{min} - I_o,\ t_{on} \le t \le T. \tag{4.42}$$

With the relationships of Equations 4.1, 4.20, 4.25, and 4.33, the capacitor current equations are reformulated as

$$i_{C,1}(t) = \frac{V_s - V_o}{L}\left[t - \frac{D}{2f}\right] \qquad (4.43)$$

and

$$i_{C,2}(t) = \frac{V_o}{L}\left[\frac{1+D}{2f} - t\right]. \qquad (4.44)$$

The solutions of Equations 4.43 and 4.44 for t when $i_C = 0$ are the zero-crossing points of the capacitor current:

$$t_1 = \frac{D}{2f}. \qquad (4.45)$$

$$t_2 = \frac{1+D}{2f}. \qquad (4.46)$$

These zero-crossing points define the time interval during which the capacitor current is positive and is thus being charged. In accordance with the charge balance principle, the charge accumulated during a periodic cycle must equal the charge dispersed during the same cycle. In other words, the capacitor must discharge the same amount it is charged each period. The charge accumulation begins at time t_1 and continues until t_2, when the capacitor current becomes zero again. These points in time also coincide with the capacitor voltage minimum (t_1) and maximum (t_2). The capacitor ripple voltage is plotted over one period in Figure 4.10 in reference to the capacitor current.

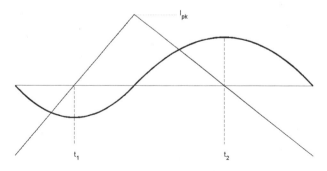

FIGURE 4.10 Capacitor current and ripple voltage.

CCM Ripple Voltage

The capacitor ripple voltage is defined as the difference between the maximum and minimum capacitor voltages:

$$\Delta V_o = v_{max} - v_{min}. \tag{4.47}$$

For a linear capacitor, this voltage difference corresponds to the amount of charge accumulated between the two voltage levels:

$$\Delta Q = C\Delta V_o. \tag{4.48}$$

The total charge accumulated is also the integral of the capacitor current over the interval $t_2 - t_1$. The result of the integration is represented by the area under $i_C(t)$, which is readily computed as the area of the triangle with base $t_2 - t_1$ and height I_{pk}:

$$\Delta Q = \frac{1}{2}(t_2 - t_1)I_{pk}. \tag{4.49}$$

The peak capacitor current is obtained from Equation 4.43 upon substitution of the relationship of Equation 4.20:

$$I_{pk} = \frac{D(V_s - V_o)}{2Lf}. \tag{4.50}$$

With Equations 4.45, 4.46, 4.49, and 4.50, the accumulated charge becomes

$$\Delta Q = \frac{1}{2}\frac{1}{2f}\frac{D(V_s - V_o)}{2Lf} = \frac{V_o(1 - D)}{8Lf^2}. \tag{4.51}$$

From Equations 4.48 and 4.51, the ripple voltage expression emerges as

$$\Delta V_o = \frac{V_o(1 - D)}{8LCf^2}. \tag{4.52}$$

The ripple voltage specification is typically stated as a fraction or as a content-percentage of the output voltage:

$$r = \frac{\Delta V_o}{V_o}. \tag{4.53}$$

Upon substitution of Equation 4.53 into Equation 4.52, the solution of Equation 4.52 for the CCM capacitance is

$$C = \frac{1-D}{8rLf^2}.$$ (4.54)

The use of a ripple specification in the computation of the CCM capacitance is illustrated in the next example.

EXAMPLE 4.5

Specify a capacitor for the buck converter in Example 4.3 for a ripple content of 1%.

Solution

Directly from Equation 4.54, the required capacitance is

$$C = \frac{1-0.36}{(8)(0.01)(160E-6)(20E3)^2} = 125 \ \mu F.$$

Conclusion

The capacitor calculation must be based on the minimum expected duty ratio in order to meet the ripple specification under all operating conditions.

EXERCISE 4.5

Derive the expressions for the CCM capacitance and the CCM peak capacitor current in terms of the λ-ratio.

Answers

$$C = \frac{1}{4\lambda Rfr}.$$ (4.55)

$$I_{pk} = \frac{I_o}{\lambda}.$$ (4.56)

THE DISCONTINUOUS CONDUCTION MODE

If the inductance in the buck converter is less than the CCM inductance, the inductor current is not sustained throughout the switching cycle but becomes zero before the end of the period. The circuit thus operates in DCM. The instantaneous inductor current and voltage are illustrated in Figure 4.11 for a DCM buck converter.

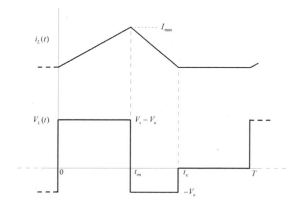

FIGURE 4.11   DCM inductor current and voltage.

DCM Inductor Current

During the on-time of SW1 in Figure 4.4, the DCM inductor current is expressed by Equation 4.14, with $I_{min} = 0$:

$$i_{L,1}(t) = \frac{V_s - V_o}{L} t, 0 \le t \le t_{on}. \tag{4.57}$$

As in CCM, the peak inductor current occurs at $t = t_{on}$:

$$I_{max} = \frac{D(V_s - V_o)}{Lf}. \tag{4.58}$$

The next conduction interval ends at t_x when the inductor current is extinguished. In this interval, $t_{on} \le t \le t_x$, the inductor current is derived from

$$i_{L,2}(t) = \int_{t_{on}}^{t} -\frac{V_o}{L} d\tau + i_{L,2}(t_{on}), \tag{4.59}$$

in which $i_{L,2}(t_{on})$ is the peak inductor current. Evaluation of the integral in Equation 4.59 and substitution of Equation 4.58 into the result yields the inductor current expression

$$i_{L,2}(t) = \frac{V_o}{L}(t_{on} - t) + \frac{D(V_s - V_o)}{Lf}, \quad t_{on} \le t \le t_x. \tag{4.60}$$

The inductor current is zero from the extinction time to the end of the switching period.

$$i_{L,3}(t) = 0, \quad t_x \le t \le T. \tag{4.61}$$

The extinction time is the solution of Equation 4.60 for t when $i_{L,2}(t) = 0$:

$$t_x = \frac{DV_s}{fV_o}. \tag{4.62}$$

With the definitions of the inductor current conduction intervals, the duty ratio is determined from the geometry of the inductor current waveform.

DCM Duty Ratio

As in CCM, the capacitor in the DCM converter is assumed to shunt all the harmonics of the inductor current and leave only the average value to flow into the load resistance. The average inductor current, which is also the load current, is computed as the area under the $i_L(t)$ curve divided by the period:

$$\frac{V_o}{R} = \frac{\frac{1}{2}t_x I_{max}}{T}. \tag{4.63}$$

Substitution of Equation 4.62 into Equation 4.63 and simplification yields

$$\frac{V_o}{R} = \frac{V_s(V_s - V_o)D^2}{2Lf V_o}. \tag{4.64}$$

The solution of Equation 4.64 for the duty ratio is

$$D = V_o \sqrt{\frac{2Lf}{RV_s(V_s - V_o)}}. \tag{4.65}$$

Unlike the duty ratio of the CCM buck converter, the duty ratio of the DCM converter is dependent upon the inductance, the switching frequency, and the load resistance. The next example illustrates an application of Equation 4.65.

EXAMPLE 4.6

Determine the duty ratio of the buck converter in Example 4.1 if the inductance is 10 μH.

Solution

Since the inductance in the circuit is less than the CCM inductance (18 μH), the converter is in DCM and, directly from Equation 4.65, the duty ratio is

$$D = 5\sqrt{\frac{(2)(10E-6)(20E3)}{(1.25)(12)(12\text{-}5)}} \approx 0.31 \text{ or } 31\%.$$

Conclusion

The duty ratio of the DCM converter is less than the duty ratio of the corresponding CCM converter. This result is consistent with the fact that a lower duty ratio results in less energy storage in the inductance. The lower energy is insufficient to sustain the inductor current throughout the entire switching period.

EXERCISE 4.6

What is the highest frequency at which the converter in Example 4.6 will operate and still be in DCM?

Answer

36 kHz.

DCM Output Voltage

A rearrangement of Equation 4.65 yields a quadratic equation for the output voltage:

$$2Lf V_o^2 + RV_oV_sD^2 - RD^2V_s^2 = 0. \tag{4.66}$$

The solution to the quadratic equation is straightforward; however, a much more direct and simple expression is available in terms of the λ-ratio relationship

$$L = \lambda L_{CCM}. \tag{4.67}$$

Substitution of Equation 4.67 into Equation 4.65 and simplification by the relationship of Equation 4.27 results in the DCM duty ratio expression

$$D = \frac{V_o}{V_s} \sqrt{\lambda}.$$ (4.68)

Equation 4.68 reveals that the DCM duty ratio is always less than the corresponding CCM duty ratio because $\lambda < 1$ in DCM. From Equation 4.68, the output voltage in terms of the λ-ratio is

$$V_o = \frac{DV_s}{\sqrt{\lambda}}.$$ (4.69)

EXAMPLE 4.7

Recompute the duty ratio for the buck converter in Example 4.6 in terms of the λ-ratio.

Solution

With an inductance of 10 µH, the λ-ratio is 10/18 = 5/9. Directly from Equation 4.68, the DCM duty ratio is

$$D = \frac{5}{12} \sqrt{5/9} \approx 0.31.$$

Conclusion

The λ-ratio greatly simplifies the duty ratio calculation for the DCM buck converter.

EXERCISE 4.7

Solve Equation 4.66 for V_o and verify that the solution yields the same numerical results as Equation 4.69 for $V_s = 15$, $D = 0.3$, $L = 20$ µH, $R = 4.2$ Ω, and $f = 20$ kHz.

Answer

$$V_o = \frac{RV_s D^2}{4Lf} \left[\sqrt{1 + \frac{8Lf}{RD^2}} - 1 \right] = 6.7 \text{ V}.$$

The λ-ratio is an indicator of the mode of operation of the buck converter. If λ is in the range $0 \leq \lambda \leq 1$, the converter is in DCM. If $\lambda > 1$, the converter is in

CCM. The next example illustrates the output voltage calculation for both modes of operation.

EXAMPLE 4.8

A buck converter has an inductance of 30 µH, a load resistance of 5 Ω, and a source voltage of 24 VDC and operates with a duty ratio of 0.5. Determine the output voltage if the PWM switching frequency is (a) 25 kHz and (b) 50 kHz.

Solution

At 25 kHz, the CCM inductance is

$$L_{CCM} = \frac{(1-0.5)(5)}{(2)(25E3)} = 50 \text{ µH}.$$

Since the actual inductance in the circuit is less than the CCM inductance, the converter operates in DCM. The λ-ratio is 30/50 = 3/5 and, from Equation 4.69, the output voltage is $V_o = (0.5)(24)\sqrt{5/3} \approx 15.5$ VDC. At 50 kHz, the CCM inductance is 25 µH; the λ-ratio is 30/25 = 6/5, and the converter operates in CCM. By Equation 4.33, the output voltage is $(0.5)(24) = 12$ VDC.

Conclusion

The mode of operation of the buck converter is readily determined by a comparison of the CCM inductance with the actual circuit inductance. The λ-ratio greatly simplifies the output voltage calculation for the DCM converter.

EXERCISE 4.8

At a switching frequency of 25 kHz, what range of load resistance will cause the converter in Example 4.8 to operate in CCM? At a frequency of 50 kHz, what range of load resistance results in DCM operation?

Answer

CCM: R < 3 Ω; DCM: R > 6 Ω.

DCM Capacitance

As in the CCM case, the capacitor current is equal to the instantaneous inductor current minus the load current. Application of the charge balance principle to the capacitor current leads to the formulation of the DCM capacitance.

DCM Capacitor Current

Subtraction of the load current from Equations 4.57, 4.60, and 4.61 yields the DCM capacitor current equations:

$$i_{C,1}(t) = \frac{V_s - V_o}{L}t - I_o, \; 0 \leq t \leq t_{on}. \tag{4.70}$$

$$i_{C,2}(t) = \frac{V_o}{L}(t_{on} - t) + \frac{D(V_s - V_o)}{Lf} - I_o, \; t_{on} \leq t \leq t_x. \tag{4.71}$$

$$i_{C,3}(t) = -I_o, \; t_x \leq t \leq T. \tag{4.72}$$

The DCM capacitor current is plotted in Figure 4.12.

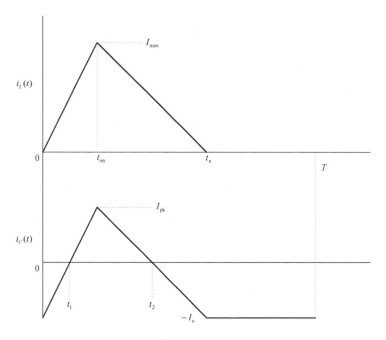

FIGURE 4.12 DCM capacitor current.

As shown in Figure 4.12, the capacitor current follows the inductor current waveform until the inductor current is extinguished. After the inductor current is zero, the capacitor must supply the load current during the remainder of the switching period. Since the capacitance is assumed to be infinite, there is no exponential decay of capacitor current after the inductor ceases to conduct; the capacitor continues to discharge into the load and maintains constant load current to the end of the period.

DCM Ripple Voltage

The charge-balance principle applied to the capacitor current yields the output ripple voltage expression of the DCM converter. The total charge is determined from the geometry of the positive portion of the DCM capacitor current illustrated in Figure 4.12.

The solutions of Equations 4.70 and 4.71 for t when $i_C(t) = 0$ are the zero-crossing points of the capacitor current:

$$t_1 = \frac{LI_o}{V_s - V_o}.$$ (4.73)

$$t_2 = t_{on} - \frac{LI_o}{V_o} + \frac{D\left(V_s - V_o\right)}{fV_o}.$$ (4.74)

The time interval during which the capacitor current is positive is

$$t_2 - t_1 = \frac{DV_s}{fV_o} - \frac{LI_oV_s}{V_o\left(V_s - V_o\right)}.$$ (4.75)

With Equations 4.27 and 4.68, the positive current conduction interval in terms of the λ-ratio is expressed as

$$t_2 - t_1 = \frac{2\sqrt{\lambda} - \lambda}{2f}.$$ (4.76)

The maximum capacitor current occurs at $t = t_{on}$. Substitution of Equation 4.20 into Equation 4.70 yields the peak capacitor expression

$$I_{pk} = \frac{D\left(V_s - V_o\right)}{Lf} - I_o.$$ (4.77)

With Equations 4.27 and 4.68 once again, the expression for the peak capacitor current is expressed in terms of the λ-ratio as

$$I_{pk} = \left(\frac{2}{\sqrt{\lambda}} - 1\right)I_o.$$ (4.78)

The accumulated charge is the integral of the capacitor current, that is, the area bounded by $i_C(t)$ and the time-axis. With Equations 4.76 and 4.78, the area under positive current, which is equal to the accumulated charge, is expressed as

$$\Delta Q = \frac{1}{2} \frac{2\sqrt{\lambda} - \lambda}{2f} \left(\frac{2}{\sqrt{\lambda}} - 1 \right) I_o = \frac{V_o \left(2 - \sqrt{\lambda} \right)^2}{4Rf}. \tag{4.79}$$

With Equations 4.48, 4.53, and 4.79, the expression for the capacitance of the DCM buck converter is formulated as

$$C = \frac{\left(2 - \sqrt{\lambda} \right)^2}{4Rfr}. \tag{4.80}$$

If $\lambda = 1$, Equation 4.80 reverts to Equation 4.55. The examples and exercises that follow illustrate some applications of Equation 4.80.

EXAMPLE 4.9

Specify a capacitor for the buck converter in Example 4.1 with a ripple specification of 1% and an inductance of (a) $L = 10\ \mu H$ and (b) $L = L_{CCM}$.

Solution

The λ-ratio for the converter when $L = 10\ \mu H$ is $10/18 = 5/9$. Directly from Equation 4.80, the required capacitance is

$$C = \frac{\left(2 - \sqrt{5/9} \right)^2}{(4)(1.25)(20E3)(0.01)} \approx 1,600\ \mu F.$$

When the inductance is L_{CCM}, the λ-ratio is unity and Equation 4.55 is applicable. The required capacitance is

$$C = \frac{1}{(4)(1.25)(20E3)(0.01)} = 1,000\ \mu F.$$

Conclusion

A significantly greater capacitance is required for the DCM converter. This result is consistent with the fact that the capacitor must sustain the load current for a longer period of time.

Exercise 4.9

Show that Equations 4.80 and 4.55 are equal when $\lambda = 1$. Determine the expression for the largest possible DCM capacitance.

Answer

$$C = \frac{1}{Rfr}.$$

Example 4.10

Design a buck converter with the following specifications:

Input Voltage	Output Voltage	Output Current	Switching Frequency	Ripple Content	λ-ratio
$30 \leq V_s \leq 42$ VDC	12 VDC	3.6 A	25 kHz	5%	2

Determine the required inductance and capacitance. Repeat the design with $\lambda = 0.5$.

Solution

With $\lambda = 2$, the converter is in CCM. The CCM inductance is

$$L_{\text{CCM}} = \frac{(1 - 12/42)(3.33)}{(2)(25E3)} \approx 48 \ \mu\text{H}.$$

The required inductance for a λ-ratio of two is $(2)(48) = 96 \ \mu$H. The required capacitance is

$$C = \frac{1}{(4)(2)(3.33)(25E3)(0.05)} \approx 30 \ \mu\text{F}.$$

With $\lambda = 0.5$, the converter is in DCM and the inductance is 24 μH. The DCM capacitance is

$$C = \frac{\left(2 - \sqrt{0.5}\right)^2}{(4)(3.33)(25E3)(0.05)} \approx 100 \ \mu\text{F}.$$

Conclusion

The CCM inductance must be based on the minimum expected CCM duty ratio. The DCM converter requires less inductance but significantly more capacitance.

EXERCISE 4.10

Redesign the converter in Example 4.10 for a load current range of $2 \leq I_o \leq 4$ A.

Answers

CCM: $L = 172$ μH, $C = 17$ μF
DCM: $L = 43$ μH, $C = 56$ μF.

PRACTICAL BUCK CONVERTERS

The previous analyses were based on ideal components: zero voltage drop across the switches and no series resistance in the inductor or in the capacitor. These assumptions facilitate the analyses but yield an unpractical circuit efficiency of 100%. In the ideal circuit, the input power is equal to the output power with no loss of energy. In practical circuits there are losses in every component. Efficiency is extremely important not only as a marketable or competitive specification but also for *thermal management*. Electronic devices that are used as switches have a nonzero voltage drop across them and generate significant amounts of heat energy that must be removed to prevent them from being destroyed. The most commonly used power devices in DC-DC converter technology are the bipolar junction transistor (BJT), the metal oxide semiconductor field-effect transistor (MOSFET), and the diode. A buck converter implemented with a BJT and diode is illustrated in Figure 4.13.

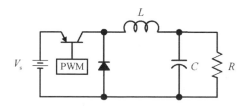

FIGURE 4.13   Practical buck converter.

CCM and DCM Duty Ratios

During the first portion of the switching period, $0 \leq t \leq t_{on}$, the transistor is switched on and the voltage difference from collector to emitter is the *saturation voltage*, V_{SAT}. The diode is reverse-biased and appears as an open circuit. When the transistor is cut off, the collector-to-emitter junction behaves as an open circuit, and the load and capacitor currents flow through the forward-biased diode back into the inductor. The voltage across the forward-biased diode is V_f. Figure 4.14 illustrates the two circuit structures during one switching period.

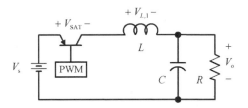

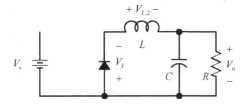

FIGURE 4.14   Switching stages of the practical buck converter.

When the transistor is on and the diode is off, application of KVL produces the inductor voltage equation

$$V_{L,1} = V_s - V_{SAT} - V_o, \ \ 0 \leq t \leq t_{on}. \tag{4.81}$$

When the diode is forward-biased and the transistor is cut off, KVL yields

$$V_{L,2} = -\left(V_o + V_f\right), \ \ t_{on} \leq t \leq T. \tag{4.82}$$

Application of the zero-average inductor-voltage principle to Equations 4.81 and 4.82 yields the expression

$$D\left(V_s - V_{SAT} - V_o\right) - \left(V_o + V_f\right)\left(1 - D\right) = 0. \tag{4.83}$$

The solution of Equation 4.83 for D is the expression for the duty ratio of the practical buck converter:

$$D = \frac{V_o + V_f}{V_s - V_{SAT} + V_f}. \tag{4.84}$$

The DCM duty ratio for the practical buck converter is

$$D = \frac{V_o + V_f}{V_s - V_{SAT} + V_f} \sqrt{\lambda}. \tag{4.85}$$

Equations 4.84 and 4.85 reveal that the duty ratio is larger in a practical buck converter than in the ideal converter.

Switching-Device Power Dissipation

The power dissipated in the transistor is found with the definition of average power applied to the collector current and the collector-to-emitter voltage:

$$P_{XTR} = \frac{1}{T} \int_0^T V_{CE}(t) i_C(t) dt. \tag{4.86}$$

In Equation 4.86 the collector-to-emitter voltage is the saturation voltage, and the collector current is equal to the inductor current in the interval $0 \le t \le t_{on}$. Since the saturation voltage is virtually constant, the transistor power dissipation is

$$P_{XTR} = V_{SAT} \left[\frac{1}{T} \int_0^{DT} i_{L,1}(t) dt \right], \tag{4.87}$$

in which the quantity in brackets is the average collector current. The diode current is equal to the inductor current during the interval $t_{on} \le t \le T$. Consequently, the diode power dissipation is

$$P_{DIO} = V_f \left[\frac{1}{T} \int_{DT}^T i_{L,2}(t) dt \right]. \tag{4.88}$$

The diode power is thus the product of the forward voltage and the average diode current. The sum of the average collector current and the average diode current is the average inductor current, which is also the load current:

$$\left[\frac{1}{T} \int_0^{DT} i_{L,1}(t) dt \right] + \left[\frac{1}{T} \int_{DT}^T i_{L,2}(t) dt \right] = I_o. \tag{4.89}$$

Equations 4.87 to 4.89 reveal that the average collector and diode currents are fractions of the load current. These fractions correspond to the duty ratios of the transistor and the diode. Consequently,

$$\frac{1}{T}\int_0^{DT} i_L(t)dt = DI_o,\tag{4.90}$$

and

$$\frac{1}{T}\int_{DT}^{T} i_L(t)dt = (1-D)I_o.\tag{4.91}$$

Substitution of the results of Equations 4.90 and 4.91 into Equations 4.87 and 4.88, respectively, yields the transistor and diode average power dissipation formulas:

$$P_{XTR} = DV_{SAT}I_o.\tag{4.92}$$

$$P_{DIO} = (1-D)V_f I_o.\tag{4.93}$$

Efficiency

The efficiency of the circuit is the ratio of the output power to the input power:

$$\eta = \frac{P_o}{P_i}.\tag{4.94}$$

The input power is the sum of the output power and the individual device losses:

$$P_i = P_o + P_{XTR} + P_{DIO}.\tag{4.95}$$

Substitution of Equation 4.95 into Equation 4.94 yields the efficiency expression

$$\eta = \frac{P_o}{P_o + P_{XTR} + P_{DIO}}.\tag{4.96}$$

Substitution of the individual power expressions into Equation 4.96 and simplification yields

$$\eta = \frac{V_o}{V_o + DV_{SAT} + (1-D)V_f}.\tag{4.97}$$

Equation 4.97 reveals that the BJT-diode converter efficiency is dependent largely upon the output voltage, and the individual voltage drops across the switching devices. To maximize efficiency, devices with low V_{SAT} and V_f are required. Additionally, converters with larger output voltages are more efficient.

The efficiency equation also reveals that for low duty ratios, the efficiency is more influenced by V_f than by V_{SAT}. Similarly, for larger duty ratios, the efficiency is affected more by V_{SAT}. For example, when the duty ratio is effectively zero, the efficiency equation becomes

$$\eta = \frac{V_o}{V_o + V_f}. \tag{4.98}$$

When the duty ratio is unity, the efficiency is

$$\eta = \frac{V_o}{V_o + V_{SAT}}. \tag{4.99}$$

The efficiency thus varies between the limits expressed by Equations 4.98 and 4.99; however, the magnitude of the limits depends upon the relative values of V_f and V_{SAT}. If $V_{SAT} > V_f$, the efficiency is lower for the larger duty ratios. If $V_f > V_{SAT}$, the efficiency is lower for the smaller duty ratios. In any case, the variation between the two limits decreases dramatically as the output voltage increases.

A DC-DC converter must be specified by its *worst-case* efficiency so that the source may be sized to meet the maximum energy demand of the circuit. The following example illustrates the application of the device dissipation and efficiency equations.

EXAMPLE 4.11

Compute the efficiency and device dissipation for the buck converter of Example 4.10 if $V_{SAT} = 0.4$ V and $V_f = 1.3$ V.

Solution

For the CCM converter ($\lambda = 2$), the duty ratio is governed by Equation 4.84, in which V_s varies between 30 and 42 VDC. Since $V_f > V_{SAT}$, the *smaller* duty ratio determines the worst-case efficiency:

$$D = \frac{12 + 1.3}{42 - 0.4 + 1.3} \approx 0.31.$$

With Equation 4.92, the transistor dissipation is computed as $P_{XTR} = (0.31)(0.4)(4.6) \approx 0.45$ W. With Equation 4.93, the diode dissipation is determined as $P_{DIO} = (0.69)(1.3)(4.6) \approx 4.23$ W. The efficiency is computed from Equation 4.97:

$$\eta = \frac{12}{12 + (0.31)(0.4) + (0.69)(0.7)} \approx 92.2\%.$$

For the DCM converter, the duty ratio is

$$D = \frac{12 + 1.3}{42 - 0.4 + 1.3}\sqrt{0.5} \approx 0.22.$$

The transistor dissipation is $P_{\text{XTR}} = (0.22)(0.4)(4.6) \approx 0.32$ W, the diode dissipation is $P_{\text{DIO}} = (0.78)(1.3)(4.6) \approx 4.65$ W, and the efficiency is

$$\eta = \frac{12}{12 + (0.22)(0.4) + (0.78)(1.3)} \approx 91.6\%.$$

Conclusion

There is clearly little difference between the efficiencies of the two converters. This result is mainly due to the relatively large input and output voltages with respect to the diode and transistor voltage drops. The DCM converter is less efficient because $V_f > V_{\text{SAT}}$; the smaller duty ratio for the transistor translates to a larger duty ratio for the diode. The choice of CCM or DCM in regards to efficiency is influenced by the relative values of V_f and V_{SAT}, the transistor and diode thermal characteristics, and the device-to-heatsink mounting configuration.

EXERCISE 4.11

a. Compute the power device dissipations and the efficiency of the buck converter in Example 4.10 with $V_{\text{SAT}} = 1.5$ and $V_f = 0.7$.
b. Compute the power device dissipations of the buck converter in Exercise 4.10 with $V_{\text{SAT}} = 0.4$ and $V_f = 1.3$.

Answer a

CCM: $P_{\text{XTR}} = 2.3$ W, $P_{\text{DIO}} = 1.4$ W, $\eta = 92\%$
DCM: $P_{\text{XTR}} = 1.7$ W, $P_{\text{DIO}} = 1.7$ W, $\eta = 93\%$.

Answer b

CCM: $P_{\text{XTR}} = 0.5$ W, $P_{\text{DIO}} = 4.6$ W
DCM: $P_{\text{XTR}} = 0.4$ W, $P_{\text{DIO}} = 4.1$ W.

Capacitor and Inductor Losses

The efficiency study would not be complete without the inclusion of the capacitor and inductor losses caused by series resistances. If these resistances are added to the circuit model, however, the analysis is completely altered: the inductor voltage and current waveforms are no longer rectangular and triangular in shape, respectively, but are exponential in form. Furthermore, the assumption of constant output voltage is compromised because of the voltage drop across the equivalent series resistance (ESR) of the capacitor, and the inductor voltage is further modified by the voltage drop across its series resistance.

An approximate approach to compute the additional losses is to continue to neglect the series resistances in the circuit model but include them in the efficiency calculation. The powers dissipated in the energy-storage devices are computed with the RMS values of the respective current waveforms. This approach results in an overestimation of the inductor/capacitor (L/C) losses but yields a more accurate and realistic circuit efficiency.

When the L/C losses are included in the efficiency calculation, the input power becomes

$$P_i = P_o + P_{XTR} + P_{DIO} + P_L + P_C, \tag{4.100}$$

in which the power dissipated in the inductor series resistance is

$$P_L = I_{L,RMS}^2 R_s, \tag{4.101}$$

and the power dissipated in the physical capacitor is

$$P_C = I_{C,RMS}^2 R_{ESR}. \tag{4.102}$$

CCM Efficiency with L/C Losses

A triangular waveform with no DC component and a peak value of X_m has an RMS value of

$$X_{\Delta,RMS} = \frac{X_m}{\sqrt{3}}. \tag{4.103}$$

With Equations 4.56 and 4.103, the RMS value of the CCM capacitor current in terms of the λ-ratio is

$$I_{C,RMS} = \frac{I_o}{\lambda\sqrt{3}}. \tag{4.104}$$

The RMS value of a triangular waveform that is level-shifted by a DC component has an RMS value that is the square-root of the sum of the squares of the RMS values of the orthogonal components:

$$X_{\Delta+\text{DC,RMS}} = \sqrt{\frac{X_m^2}{3} + X_{\text{DC}}^2} . \tag{4.105}$$

With Equation 4.56 again and Equation 4.105, the RMS value of the CCM inductor current in terms of the λ-ratio is

$$I_{L,\text{RMS}} = I_o \sqrt{\frac{1}{3\lambda^2} + 1} . \tag{4.106}$$

Equations 4.104 and 4.106 determine the CCM L/C losses as follows:

$$P_L = \left(\frac{1}{3\lambda^2} + 1 \right) I_o^2 R_s . \tag{4.107}$$

$$P_C = \frac{1}{3\lambda^2} I_o^2 R_{\text{ESR}} . \tag{4.108}$$

Equations 4.107 and 4.108 determine the CCM efficiency with L/C losses to be

$$\eta = \frac{V_o}{V_o + DV_{\text{SAT}} + (1-D)V_f + \dfrac{I_o}{3\lambda^2}\left(R_s + R_{\text{ESR}} \right) + I_o R_s} . \tag{4.109}$$

Equation 4.109 reveals that the efficiency is dependent upon additional voltage drops. These voltages are related to the average voltages across R_s and R_{ESR}. The equation further reveals that it is possible to reduce the L/C losses with larger λ-ratios (larger inductance values) as long as the inductor series resistance, R_s, is not subsequently increased.

DCM Efficiency with L/C Losses

Equation 4.103 also applies to a triangular waveform with a maximum of X_m and a minimum of zero. If the base of the triangle is less than the period of the waveform, the RMS value is

$$X_{\text{RMS}} = X_m \sqrt{\frac{t_b}{3T}} , \tag{4.110}$$

in which t_b is the length of the base. The peak DCM inductor current is determined by the addition of I_o to Equation 4.78:

$$I_{\max} = \frac{2I_o}{\sqrt{\lambda}}. \tag{4.111}$$

As illustrated in Figure 4.12, the base of the triangular DCM inductor current waveform is the extinction time, t_x. With Equations 4.62 and 4.68, the extinction time is expressed in terms of the λ-ratio as

$$t_x = \frac{\sqrt{\lambda}}{f}. \tag{4.112}$$

With Equations 4.1, 4.110, 4.111, and 4.112, the RMS value of the DCM inductor current as a function of the λ-ratio is

$$I_{L,\mathrm{RMS}} = \frac{2I_o}{\sqrt{3}\lambda^{1/4}}. \tag{4.113}$$

The RMS DCM capacitor current is determined from the piece-wise waveform geometry illustrated in Figure 4.12 and the respective duty ratio segments:

$$I_{C,\mathrm{RMS}} = \sqrt{\frac{I_{pk}^2}{3}\frac{t_x}{T} + I_o^2 \frac{T - t_x}{T}}. \tag{4.114}$$

With Equations 4.1, 4.78, 4.112, and 4.114, the RMS DCM capacitor current in terms of the λ-ratio is

$$I_{C,\mathrm{RMS}} = \frac{I_o}{\sqrt{3}\lambda^{1/4}}\sqrt{4 - 2\lambda - \sqrt{\lambda}}. \tag{4.115}$$

Equations 4.113 and 4.115 determine the DCM L/C losses as follows:

$$P_L = \frac{4I_o^2}{3\sqrt{\lambda}}R_s. \tag{4.116}$$

$$P_C = \frac{I_o^2}{3\sqrt{\lambda}}\left(4 - 2\lambda - \sqrt{\lambda}\right)R_{\mathrm{ESR}}. \tag{4.117}$$

The next example illustrates how the L/C losses are incorporated into the efficiency calculations.

EXAMPLE **4.12**

Compute the L/C losses and the circuit efficiency of the buck converter in Example 4.11 with $R_s = 0.25 \ \Omega$ and $R_{\text{ESR}} = 0.1 \ \Omega$.

Solution

From Equations 4.104 and 4.106, the RMS CCM inductor and capacitor currents are

$$I_{L,\text{RMS}} = (3.6)\sqrt{\frac{1}{(3)(2)} + 1} = 3.75 \ \text{A}$$

and

$$I_{C,\text{RMS}} = \frac{3.6}{(2)\sqrt{3}} = 1.04 \ \text{A}.$$

With Equations 4.101 and 4.102, the CCM L/C losses are determined as

$$P_L = (1.36)^2(0.25) = 3.51 \ \text{W}$$

and

$$P_C = (1.04)^2(0.1) = 0.11 \ \text{W}.$$

The efficiency is computed with Equations 4.94 and 4.100 or with Equation 4.109:

$$\eta = \frac{12}{12 + (0.31)(0.4) + (0.69)(1.3) + \dfrac{3.6}{(3)(2)^2}\big(0.25 + 0.1\big) + (3.6)(0.25)} = 85.6\%.$$

The RMS DCM inductor and capacitor currents are determined from Equations 4.113 and 4.115 to be

$$I_{L,\text{RMS}} = \frac{(2)(3.6)}{\sqrt{3}(0.5)^{0.25}} = 4.94 \ \text{A}$$

and

$$I_{C,RMS} = \frac{(3.6)}{\sqrt{3}(0.5)^{0.25}}\sqrt{4 - (2)(0.5) - \sqrt{0.5}} = 3.74 \ \text{A}.$$

With Equations 4.101 and 4.102 again, the DCM L/C losses are determined as

$$P_L = (4.94)^2 (0.25) = 6.1 \text{ W}$$

and

$$P_C = (3.74)^2 (0.1) = 1.4 \text{ W}.$$

The DCM efficiency is

$$\eta = \frac{43.2}{43.2 + 0.32 + 3.65 + 6.11 + 1.4} = 79\%.$$

Conclusion

The L/C losses and efficiency may be calculated directly with their respective equations or calculated, as in this example, in separate steps that begin with the RMS current values. The DCM converter is less efficient because the smaller λ-ratio results in larger RMS current values.

EXERCISE 4.12

To increase the efficiency of the CCM converter in Example 4.12, is it more effective to use a larger inductance or to use an inductance with a smaller series resistance? Compute the efficiency of the CCM converter for (a) $\lambda = 10$ and $R_s = 0.25$ and (b) $\lambda = 2$ and $R_s = 0.1$.

Answers

(a) $\eta = 86.2\%$ and (b) $\eta = 89.3\%$.

CHAPTER SUMMARY

The buck converter is an effective circuit solution to the problems associated with the supply of a lower DC voltage from a higher voltage DC source. The efficiency issue is addressed by the saturated switching of the main power switch, and the regulation problem is solved by PWM.

The λ-ratio is a new addition to the theory of DC-DC converters. The ratio provides a means to determine the mode of operation and clearly shows how an increase or decrease in inductance affects the design issues of the circuit. In addition, the λ-ratio greatly simplifies the analysis of the DCM converter.

The ideal CCM buck converter equations are summarized as follows:

$$D = \frac{V_o}{V_s},$$

$$L_{CCM} = \frac{(1-D)R}{2f},$$

$$C = \frac{1}{4\lambda R f r}.$$

The ideal DCM buck converter equations are summarized as follows:

$$D = \frac{V_o}{V_s}\sqrt{\lambda},$$

$$C = \frac{\left(2-\sqrt{\lambda}\right)^2}{4Rfr}.$$

Of the efficiency expressions developed in the chapter, the choice of which equation to apply depends on the desired computational accuracy and the type of semiconductor devices used in the circuit design.

THE MATLAB TOOLBOX

ON THE CD

The functions listed below are located in the Toolbox subfolder of Chapter 4 on the CD-ROM.

LBUCK

Function *lbuck.m* computes the minimum inductance required for a buck converter from the duty ratio, λ-ratio, load resistance, and PWM switching frequency.

Syntax

```
L = lbuck(D, lambda, R, f)
```

CBUCK

Function *cbuck.m* computes the minimum required capacitance for a buck converter from the ripple, λ-ratio, load resistance, and PWM switching frequency.

Syntax

```
C = cbuck(r, lambda, R, f)
```

PROBLEMS

For the design problems, compute the minimum inductance and capacitance that meets the circuit specification. For the efficiency problems, compute the worst-case efficiency.

Problem 1

Design a CCM buck converter to output 12 VDC at 50 W with 2% ripple from a source that varies between 30 and 42 VDC with a PWM switching frequency of 25 kHz.

Problem 2

Design a DCM buck converter to output 12 VDC at 50 W with 2% ripple from a source that varies between 30 and 42 VDC with a PWM switching frequency of 25 kHz.

Problem 3

Design a CCM buck converter to output 5 VDC at 4 to 6 A with 1% ripple from a source that varies between 10 and 14 VDC with a PWM switching frequency of 20 kHz.

Problem 4

Design a DCM buck converter to output 5 VDC at 4 to 6 A with 1% ripple from a source that varies between 10 and 14 VDC with a PWM switching frequency of 20 kHz.

Problem 5

Repeat Problem 1 for a buck converter implemented with a BJT and diode with $V_{SAT} = 0.5$ V and $V_f = 0.7$. Compute the circuit efficiency.

Problem 6

Repeat Problem 2 for a buck converter implemented with a BJT and diode with $V_{SAT} = 0.5$ V and $V_f = 0.7$. Compute the circuit efficiency.

Problem 7

Repeat Problem 3 for a buck converter implemented with a BJT and diode with $V_{SAT} = 0.5$ V and $V_f = 0.7$. Compute the circuit efficiency.

Problem 8

Repeat Problem 4 for a buck converter implemented with a BJT and diode with $V_{SAT} = 0.5$ V and $V_f = 0.7$. Compute the circuit efficiency.

Problem 9

Repeat Problem 1 for a buck converter implemented with two MOSFETS with on-resistances of 0.01 Ω, an inductor with a series resistance of 0.1 Ω, and a capacitor with an ESR of 0.05 Ω. Compute the circuit efficiency.

Problem 10

Repeat Problem 2 for a buck converter implemented with two MOSFETS with on-resistances of 0.01 Ω, an inductor with a series resistance of 0.1 Ω, and a capacitor with an ESR of 0.05 Ω. Compute the circuit efficiency.

Problem 11

Repeat Problem 3 for a buck converter implemented with two MOSFETS with on-resistances of 0.01 Ω, an inductor with a series resistance of 0.1 Ω, and a capacitor with an ESR of 0.05 Ω. Compute the circuit efficiency.

Problem 12

Repeat Problem 4 for a buck converter implemented with two MOSFETS with on-resistances of 0.01 Ω, an inductor with a series resistance of 0.1 Ω, and a capacitor with an ESR of 0.05 Ω. Compute the circuit efficiency.

5 General Theory of Two-Level DC-DC Converters: Boost and Buck/Boost Converters

In This Chapter

INTRODUCTION

The design equations developed for the buck converter in Chapter 3 are a subset of a general set of equations that are applicable to a great variety of DC-DC converters. These types of converters are called *two-level* DC-DC converters because of the two voltage levels of the inductor voltage waveform. In this chapter the principles of analysis applied to the buck converter are generalized as a theory from which the design equations for other converters are formulated. The general theory is developed and applied to the boost and buck/boost converters, both in CCM and DCM operation.

The terms *buck* and *boost* come from electrical transformer jargon. A bucking, or buck, transformer is a step-down transformer that outputs a lower secondary voltage than the primary source. The rated secondary current is greater than the primary current by a factor of the ratio of the primary to secondary voltages. A boost transformer steps the source voltage up to a higher voltage but requires a primary current that is greater than the secondary current by a factor of the ratio of the secondary to primary voltages. These voltage and current relationships are also true for the buck and boost converters, except that the voltages and currents are DC rather than AC. Indeed, DC-DC converters are often called "DC transformers." As the name implies, the buck/boost converter can output voltage magnitudes that are greater or less than the source.

GENERAL THEORY OF CCM CONVERTERS

The primary energy-transfer element in DC-DC converters is the inductor; capacitors are seldom used. The general theory of CCM converters begins with an analysis of the voltages across the inductor throughout the switching period. From the inductor voltage waveform, the general expression for the duty ratio follows from the volt-second balance principle.

The analysis continues with a formulation of the inductor current from which the expression for the CCM inductance is determined. The capacitor current is then formulated, along with the output ripple voltage and, finally, the expression for the CCM capacitance.

Generalized CCM Duty Ratio

The steady-state inductor voltage waveform of a general CCM converter is shown in Figure 5.1.

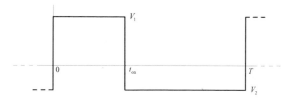

FIGURE 5.1 General CCM inductor voltage waveform.

As indicated in Figure 5.1, the voltage across the inductor is

$$V_{L,1}(t) = V_1, \ 0 \le t \le t_{on}. \tag{5.1}$$

$$V_{L,2}(t) = V_2, \ t_{on} < t \le T. \tag{5.2}$$

Application of the zero-average principle produces the identity

$$DV_1 + (1 - D)V_2 = 0. \tag{5.3}$$

The solution of Equation 5.3 for the duty ratio is

$$D_{\text{CCM}} = \frac{V_2}{V_2 - V_1}. \tag{5.4}$$

Equation 5.4 is the general duty ratio expression for CCM converters in terms of the on-time and off-time voltages across the inductor. Example 5.1 illustrates an application of the generalized duty ratio expression.

EXAMPLE 5.1

Derive the CCM duty ratio expression for the buck converter from Equation 5.4.

Solution

The on-time voltage across the inductance of the CCM buck converter is $V_1 = V_s - V_o$; the off-time voltage is $V_2 = -V_o$. Substitution of the on- and off-time voltage expressions into Equation (5.4) results in

$$D = \frac{-V_o}{-V_o - (V_s - V_o)} = \frac{-V_o}{-V_o - V_s + V_o} = \frac{-V_o}{-V_s} = \frac{V_o}{V_s}.$$

Conclusion

Equation 5.4 yields the expression for the CCM buck converter duty ratio directly without need of further analysis.

EXERCISE 5.1

A certain CCM converter has an inductor on-time voltage of 24 V and an off-time voltage of negative −12 V. Determine the duty ratio of the converter.

Answer

$$D = \frac{1}{3}.$$

Average Inductor Current and Load Current

Application of the inductor current integral to Equations 5.1 and 5.2 results in the expressions for the general CCM inductor current:

$$i_{L,1}(t) = \frac{V_1}{L} t + I_{\min}, \ \ 0 \le t \le t_{\text{on}}. \tag{5.5}$$

$$i_{L,2}(t) = \frac{V_2}{L}(t - t_{\text{on}}) + I_{\max}, \ \ t_{\text{on}} \le t \le T. \tag{5.6}$$

The generalized CCM inductor current is illustrated in Figure 5.2.

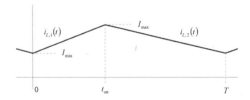

FIGURE 5.2 Generalized CCM inductor current.

If the capacitor shunts all the harmonics of $i_L(t)$, as assumed in Chapter 4, the load current is equal to the average inductor current. However, in some converters the inductor is disconnected from the R-C load during a portion of the switching cycle. In these converters, only a fraction of the average inductor current flows into the load. This fraction is defined as

$$D_{\text{o}} = \frac{t_{\text{o}}}{T}, \tag{5.7}$$

in which t_{o} is the time interval during which the inductor is load-connected. If the inductor is always connected to the load, as in the buck converter, the load-connected time interval is the switching period, and the load current is the average inductor current. If the inductor is not always load-connected, then the load current is

$$I_{\text{o}} = D_{\text{o}} I_{\text{avg}}. \tag{5.8}$$

If the inductance is load-connected only while the power switch is on, then $t_o = t_{on}$ and

$$D_o = D. \tag{5.9}$$

If the inductance is load-connected only during the power switch off-time, then $t_o = T - t_{on}$ and

$$D_o = 1 - D. \tag{5.10}$$

The *load-connected duty ratio*, in accordance with Equation 5.8, is the ratio of the load current to the average inductor current:

$$D_o = \frac{I_o}{I_{avg}}. \tag{5.11}$$

The average inductor current is computed from the geometry of the inductor current waveform as illustrated in Figure 5.2. As indicated in the figure, the peak inductor current occurs at $t = t_{on}$. With the on-time expressed as $t_{on} = \dfrac{D}{f}$, the peak inductor current is formulated as

$$I_{max} = \frac{DV_1}{Lf} + I_{min}. \tag{5.12}$$

The sum of the rectangular area and the triangular area of the inductor current waveform illustrated in Figure 5.2 is computed as

$$A = \frac{1}{2}T\left(\frac{DV_1}{Lf}\right) + TI_{min}. \tag{5.13}$$

The average inductor current is the ratio of the area expressed by Equation 5.13 to the PWM switching period:

$$I_{avg} = \frac{DV_1}{2Lf} + I_{min}. \tag{5.14}$$

The product of the average inductor current and the load-connected duty ratio results in the load current:

$$I_o = \frac{DD_oV_1}{2Lf} + D_oI_{min}. \tag{5.15}$$

Generalized CCM Inductance

The solution of Equation 5.15 for L when $I_{min} = 0$ is the minimum inductance required for CCM operation:

$$L_{CCM} = \frac{DD_o V_1 R}{2 f V_o}.$$ (5.16)

Example 5.2 illustrates an application of Equation 5.16.

EXAMPLE 5.2

Derive the expression for the CCM inductance of the buck converter from Equation 5.16.

Solution

As stated previously, the inductor is always connected to the load in the buck converter. The load-connected duty ratio, therefore, is $D_o = 1$. With the on-time voltage from Example 5.1 and a duty ratio of $D = V_o/V_s$, the CCM inductance is computed as

$$L_{CCM} = \frac{\frac{V_o}{V_s}(1)(V_s - V_o)R}{2 f V_o} = \frac{\left(1 - \frac{V_o}{V_s}\right)R}{2f} = \frac{(1-D)R}{2f}.$$

Conclusion

Equation 5.16 directly yields the CCM inductance for the buck converter.

EXERCISE 5.2

A certain CCM converter has an inductor on-time voltage of $V_1 = V_s$ and a load-connected duty ratio of $D_o = 1 - D$. Determine the expression for the CCM inductance.

Answer

$$L_{CCM} = \frac{D(1-D)V_s R}{2 V_o f}.$$

CCM Capacitance

When the inductor is load-connected, the capacitor current is equal to the inductor current minus the load current. When the inductor is disconnected from the load, the capacitor must supply the load current. The capacitor current is thus related to the load-connected duty ratio. The capacitor currents are illustrated in Figure 5.3 along with their corresponding load-connected duty ratios.

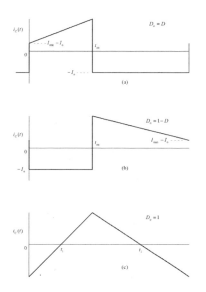

FIGURE 5.3   CCM capacitor currents and load-connected duty ratios.

When $D_o = D$, the CCM capacitor current is

$$i_{C,1}(t) = \frac{V_1}{L}t + I_{\min} - I_o, \ 0 \le t \le t_{\text{on}}. \tag{5.17}$$

$$i_{C,2}(t) = -I_o, \ t_{\text{on}} \le t \le T. \tag{5.18}$$

If $D_o = 1 - D$, the capacitor current is

$$i_{C,1}(t) = -I_o, \ 0 \le t \le t_{\text{on}}. \tag{5.19}$$

$$i_{C,2}(t) = \frac{V_2}{L}(t - t_{\text{on}}) + I_{\max} - I_o, \ t_{\text{on}} \le t \le T. \tag{5.20}$$

When $D_o = 1$, the capacitor current is expressed by Equations 5.17 and 5.20.

Capacitor Charge Balance

In accordance with the charge-balance principle, the integral of the capacitor current over the switching period is zero. Because the integral represents area *and* charge, the areas above and below the time-axes in Figure 5.3 must have equal magnitudes. When $D_o = D$, the magnitude of area is readily calculated as the area of the rectangle in the interval between T and t_{on}. Application of the charge-balance principle to the waveform in Figure 5.3a results in

$$C\Delta V_o = (T - t_{on})I_o. \tag{5.21}$$

In terms of the switching frequency and duty ratio, Equation 5.21 is reformulated as

$$C\Delta V_o = \frac{(1-D)V_o}{Rf}. \tag{5.22}$$

The solution of Equation 5.22 for the ripple voltage is

$$\Delta V_o = \frac{(1-D)V_o}{RCf}. \tag{5.23}$$

With the definition of ripple voltage from Chapter 4, the solution of Equation 5.23 for C yields the expression for the CCM capacitance when $D_o = D$:

$$C = \frac{1-D}{Rfr}. \tag{5.24}$$

Equation 5.24 indicates that larger capacitances are required for smaller duty ratios when the inductor is load-connected during the on-time. This conclusion is consistent with the operation of the circuit since the capacitor must sustain the load current for a longer time interval.

The charge-balance principle applied to the waveform in Figure 5.3b produces

$$C\Delta V_o = I_o t_{on}. \tag{5.25}$$

In terms of D and f, Equation 5.25 is expressed as

$$C\Delta V_o = \frac{DV_o}{Rf}. \tag{5.26}$$

With the definition of ripple again, the CCM capacitance when $D_o = 1 - D$ is

$$C = \frac{D}{Rfr}. \tag{5.27}$$

When $D_o = 1$, the load current expressed by Equation 5.15 becomes

$$I_o = \frac{DV_1}{2Lf} + I_{min}. \tag{5.28}$$

As illustrated in Figure 5.3c, the accumulated charge is equal to the area of the triangle above the time axis. To compute the area, the zero-crossing points are required and are readily obtained by a reformulation of Equations 5.17 and 5.20 in terms of the duty ratio and frequency:

$$i_{C,1}(t) = \frac{V_1}{L}\left[t - \frac{D}{2f} \right], \quad 0 \le t \le t_{on}. \tag{5.29}$$

$$i_{C,2}(t) = \frac{V_2}{L}\left[t - \frac{1+D}{2f} \right], \quad t_{on} \le t \le T. \tag{5.30}$$

The solutions of Equations 5.29 and 5.30 for $i_C(t) = 0$ yield the zero-crossing points. The difference between the zero-crossing times is the length of the base of the triangle:

$$t_2 - t_1 = \frac{1}{2f}. \tag{5.31}$$

The height of the triangle is the peak capacitor current that occurs at the same time as the maximum inductor current. Substitution of $t = t_{on}$ into Equation 5.29 yields

$$I_{pk} = \frac{DV_1}{2Lf}. \tag{5.32}$$

With Equations 5.31 and 5.32, the accumulated charge is computed as

$$\Delta Q = \frac{1}{2}\frac{1}{2f}\frac{DV_1}{2Lf} = \frac{DV_1}{8Lf^2}. \tag{5.33}$$

The charge-balance principle, along with the definition of ripple, applied to Equation 5.33 yields the CCM capacitance for a unity load-connected duty ratio:

$$C = \frac{DV_1}{8rV_oLf^2}. \tag{5.34}$$

EXAMPLE 5.3

Derive the expression for the CCM capacitance of the buck converter from Equation 5.34.

Solution

For the CCM buck converter, $V_1 = V_s - V_o$ and $D = V_o/V_s$. Substitution of these expressions into Equation 5.34 and simplification yields

$$C = \frac{D(V_s - V_o)}{8rV_oLf^2} = \frac{DV_o\left(\frac{V_s}{V_o} - 1\right)}{8rV_oLf^2} = \frac{D\left(\frac{1}{D} - 1\right)}{8rLf^2} = \frac{1-D}{8rLf^2}.$$

Conclusion

Equation 5.34 yields directly the CCM capacitance for the buck converter.

EXERCISE 5.3

Determine the CCM capacitance for the converter in Exercise 5.2.

Answer

$$C = \frac{D}{Rfr}.$$

The CCM Boost Converter

As stated earlier, the boost converter outputs a voltage that is greater than the input voltage. The functional schematic of the boost converter is shown in Figure 5.4.

The equivalent circuits of the boost converter, in accordance with the switch closures, are illustrated in Figure 5.5.

As indicated in Figure 5.5a, the inductor is switched directly across the DC source during the on-time; thus $V_1 = V_s$. When SW1 is open and SW2 is closed, as illustrated in Figure 5.5b, the inductor is load-connected and the voltage across the

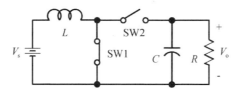

FIGURE 5.4   Boost converter.

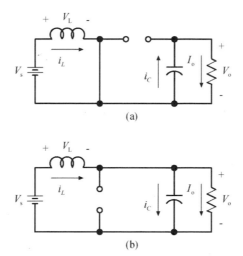

(a)

(b)

FIGURE 5.5   Equivalent circuits of the boost converter.

inductor is $V_2 = V_s - V_o$. Substitution of these voltage expressions into Equation 5.4 yields the CCM duty ratio of the boost converter:

$$D = 1 - \frac{V_s}{V_o}. \tag{5.35}$$

The solution of Equation 5.35 for the ratio of the output-to-input voltages results in the transfer function of the circuit:

$$\frac{V_o}{V_s} = \frac{1}{1-D}. \tag{5.36}$$

Although Equation 5.36 indicates a possible limitless output voltage, the properties of the physical components impose an upper bound on the output.

CCM Inductance

The boost converter inductance is load-connected during the off-time of SW1. The load-connected duty ratio is therefore $1 - D$. Substitution of Equations 5.10, 5.35, and 5.36 into Equation 5.16 yields the CCM inductance for the boost converter:

$$L_{CCM} = \frac{D(1-D)^2 R}{2f}. \tag{5.37}$$

As stated in Chapter 4, the CCM inductance specified for a converter must be the largest value computed under varying circuit conditions. For the buck converter, the largest CCM inductance occurs at maximum load resistance and minimum duty ratio. For the boost converter, the largest inductance occurs with maximum resistance as well; however, the minimum duty ratio does not provide the largest inductance. Differentiation of Equation 5.37 with respect to the duty ratio yields the expression

$$\frac{dL_{CCM}}{dD} = \frac{\left(1-4D+3D^2\right)R}{2f}. \tag{5.38}$$

The maximum inductance is determined with Equation 5.38 set to zero:

$$1-4D+3D^2 = 0. \tag{5.39}$$

The solutions to the quadratic equation are

$$D = \begin{cases} 1/3 \\ 1 \end{cases}.$$

A duty ratio of unity is not practical, nor does it yield maximum inductance. The maximum inductance therefore occurs at a duty ratio of one-third, as illustrated in Figure 5.6, in which the normalized CCM inductance is plotted versus duty ratio.

According to Figure 5.6 and the applicable solution to Equation 5.39, the largest required CCM inductance occurs at the operating duty ratio that is closest to one-third.

Example 5.4 illustrates the CCM inductance calculation for the boost converter under variable circuit conditions.

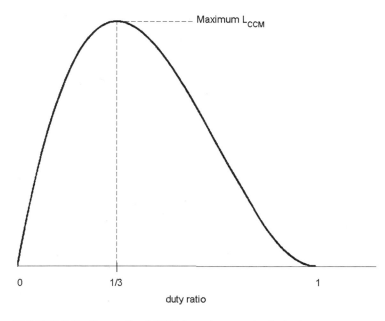

FIGURE 5.6   Normalized CCM boost converter inductance.

EXAMPLE 5.4

A boost converter has an output voltage of 48 VDC, with an input voltage that varies from 10 to 14 VDC. The load current varies between 3 and 6 A. Compute the CCM inductance if the PWM switching frequency is 5 kHz.

Solution

The maximum load resistance occurs at minimum load current: $R = 48/3 = 16\ \Omega$. The duty ratio at an input voltage of 10 VDC is $D = 1 - 10/48 = 0.79$, or 79%. The duty ratio at 14 VDC is $D = 1 - 14/48 = 0.71$, or 71%. Since 71% is closer to 33%, the CCM inductance must be based on the 71% duty ratio to ensure CCM operation under all circuit variations. The CCM inductance therefore is

$$L_{\mathrm{CCM}} = \frac{(0.71)(1-0.71)(3)}{(2)(5E3)} = 18\ \mu\mathrm{H}.$$

Conclusion

If the CCM inductance is based on the 79% duty ratio, the approximate value is 10 μH. It is evident that a small variation in duty ratio can result in a large variation of CCM inductance. If the inductance is not at least 18 µH for this example, the converter will not always operate in CCM; consequently, the output voltage will not always be 48 VDC.

EXERCISE 5.4

Determine the expression for the maximum CCM inductance for a boost converter under all possible operating conditions.

Answer

$$L_{max} = \frac{2 R_{max}}{27 f}.$$

CCM Boost Converter Capacitance

Since the load-connected duty ratio of the boost converter is $D_o = 1 - D$, the CCM capacitance is specified by Equation 5.27.

The CCM Buck/Boost Converter

As stated in the introduction, the buck/boost converter has the capability to provide output voltage magnitudes that are less than or greater than the input voltage. If the negative terminal of the source is the reference node, the polarity of the output voltage is opposite the polarity of the source voltage. The functional schematic of the buck/boost converter is shown in Figure 5.7.

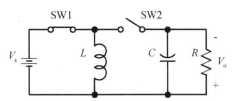

FIGURE 5.7 Buck/boost converter.

The equivalent circuits of the buck/boost converter in accordance with the switch closures are illustrated in Figure 5.8.

As illustrated in Figure 5.8, the voltage across the inductor during the on-time is, as the boost converter, $V_1 = V_s$. During the off-time, the inductor voltage is

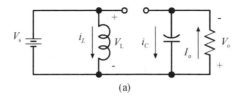

(a)

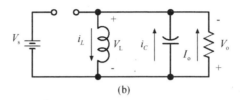

(b)

FIGURE 5.8   Equivalent circuits of the buck/boost converter.

$V_2 = -V_o$. Substitution of these voltages into Equation 5.4 yields the CCM duty ratio of the buck/boost converter:

$$D = \frac{V_o}{V_o + V_s}. \tag{5.40}$$

From Equation 5.40, the transfer function of the buck/boost converter is determined as

$$\frac{V_o}{V_s} = \frac{D}{1-D}. \tag{5.41}$$

Equation 5.41 indicates that if the duty ratio is less than one-half, the magnitude of the output voltage is less than the input voltage; if the duty ratio is greater than one-half, the output voltage is greater than the input voltage. At a duty ratio of exactly one-half, the input and output voltage magnitudes are equal, but with opposite polarities. As with the boost converter, the output voltage is bounded by the physical limitations of the components.

CCM Inductance

As with the boost converter, the inductance of the buck/boost converter is load-connected during the off-time of SW1. Substitution of Equations 5.10, 5.40, and 5.41 into Equation 5.16 yields the CCM inductance for the buck/boost converter:

$$L_{CCM} = \frac{(1-D)^2 R}{2f}. \tag{5.42}$$

As with the buck converter, the CCM inductance for the buck/boost converter is based on maximum load resistance and minimum duty ratio. The next example illustrates the CCM inductance computation for the buck/boost converter.

EXAMPLE 5.5

A buck/boost converter supplies a 28-VDC load from a source that varies between 24 and 32 VDC. The load current varies from 7 to 10 A. Determine the CCM inductance if the PWM switching frequency is 10 kHz.

Solution

The load resistance varies between 28/10 = 2.8 Ω and 28/7 = 4 Ω. The duty ratio varies between 28/(32 + 28) = 0.47 and 28/(24 + 28) = 0.55. The required CCM inductance is

$$L_{CCM} = \frac{(1-0.47)^2 (4)}{(2)(10E3)} = 57 \; \mu H.$$

Conclusion

To maintain CCM operation under the specified variations in load current and input voltage, the CCM inductance must larger than 57 μH.

EXERCISE 5.5

A 100 μH inductance is used as the CCM inductance of a buck/boost converter with an output voltage of 144 VDC and an expected input voltage variation of 120 to 162 VDC. The expected output current variation is 5 to 10 amperes. What is the minimum PWM switching frequency that ensures CCM operation?

Answer

40 kHz.

CCM Capacitance

The load-connected duty ratio of the buck/boost converter is the same as the boost converter; hence, the CCM capacitance specification is also given by Equation 5.27.

GENERAL THEORY OF DCM CONVERTERS

As discussed in Chapter 4, the voltage across and the current through the energy-storage inductance is zero over a portion of the switching period when a converter is in DCM. These voltage and current waveforms of the general DCM converter are illustrated in Figure 5.9.

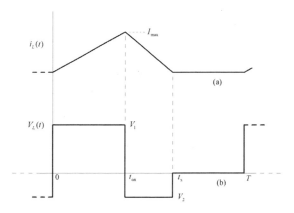

FIGURE 5.9   Inductor voltage and current waveforms of the generalized DCM converter.

As indicated in Figure 5.9, the voltage across the inductor during the on-time is the same as Equation 5.1. The voltage during the off-time is

$$V_{L,2}(t) = \begin{cases} V_2, & t_{on} \leq t \leq t_x \\ 0, & t_x \leq t \leq T \end{cases}.$$ (5.43)

Application of the zero-average principle to the inductor voltage waveform produces an expression for the extinction time:

$$t_x = \frac{D}{f}\left(1 - \frac{V_1}{V_2}\right).$$ (5.44)

Generalized DCM Duty Ratio

As demonstrated by Equation 5.15, the relationship between the average inductor current and the load current is dependent upon the duty ratio and the load-connected duty ratio. It is possible, therefore, to extract a general DCM duty ratio expression from that relationship. The average value of the general DCM inductor current is readily computed with the base and height of the triangular waveform in Figure 5.9a. The height of the triangle is the maximum inductor current expressed by Equation 5.12 with $I_{min} = 0$:

$$I_{max} = \frac{DV_1}{Lf}. \tag{5.45}$$

The base of the triangle is the extinction time. The average DCM inductor current is thus

$$I_{avg} = \frac{D^2V_1}{2Lf}\left(1 - \frac{V_1}{V_2}\right). \tag{5.46}$$

In accordance with Equation 5.8, the output current is the product of the average inductor current and the load-connected duty ratio:

$$\frac{V_o}{R} = \frac{D_o D^2 V_1}{2Lf}\left(1 - \frac{V_1}{V_2}\right). \tag{5.47}$$

The solution of Equation 5.47 for D yields the expression for the generalized DCM duty ratio:

$$D = \sqrt{\frac{2Lf\,V_o V_2}{RD_o V_1\left(V_2 - V_1\right)}}. \tag{5.48}$$

EXAMPLE 5.6

Derive the expression for the DCM duty ratio of the buck converter from Equation 5.48.

Solution

Substitution of the buck converter inductor voltages from Example 5.1 yields

$$D = \sqrt{\frac{2Lf\,V_o(-V_o)}{R\left(V_s - V_o\right)\left(-V_s\right)}} = V_o\sqrt{\frac{2Lf}{RV_s\left(V_s - V_o\right)}}.$$

Conclusion

Equation 5.48 yields directly the DCM duty ratio of the buck converter.

EXERCISE 5.6

Derive the expression for the DCM duty ratio of a DC-DC converter with inductor voltages $V_1 = V_s$ and $V_2 = -V_o$. The load-connected duty ratio is $D_o = 1 - D$, in which D is expressed by Equation 5.40.

Answer

$$D = \frac{V_o}{V_s}\sqrt{\frac{2Lf}{R}}.$$

Generalized Lambda Ratio and DCM Inductance

The generalized λ-ratio is formed by the ratio of the circuit inductance to the generalized CCM inductance as expressed by Equation 5.16:

$$\lambda = \frac{2Lf V_o}{DD_o V_1 R}. \tag{5.49}$$

Since the λ-ratio is based on the CCM inductance, the duty ratio in Equation 5.49 is the CCM duty ratio. Substitution of Equation 5.4 into Equation 5.49 results in

$$\lambda = \frac{2Lf V_o \left(V_2 - V_1\right)}{D_o V_1 V_2 R}. \tag{5.50}$$

The solution of Equation 5.50 for L is the generalized DCM circuit inductance:

$$L = \frac{D_o V_1 V_2 R}{2f V_o \left(V_2 - V_1\right)}\lambda. \tag{5.51}$$

Example 5.7 illustrates the application of Equation 5.51 to the DCM buck converter.

EXAMPLE 5.7

Derive the DCM duty ratio of the buck converter in terms of the λ-ratio from Equation 5.51 and the duty ratio expression in Example 5.6.

Solution

Substitution of the buck converter inductor voltages and a load-connected duty ratio of unity into Equation 5.51 results in a circuit inductance of

$$L = \frac{(V_s - V_o)R}{2fV_s}\lambda.$$

Substitution of the inductance into the expression from Example 5.6 results in

$$D = V_o\sqrt{\frac{\frac{(V_s - V_o)R}{V_s}\lambda}{RV_s(V_s - V_o)}} = V_o\sqrt{\frac{\lambda}{V_s^2}} = \frac{V_o}{V_s}\sqrt{\lambda}.$$

Conclusion

The generalized λ-ratio and inductance equations yield directly the DCM duty ratio of the buck converter as a function of λ.

EXERCISE 5.7

Express the DCM duty ratio of the converter in Exercise 5.6 in terms of the λ-ratio.

Answer

$$D = \frac{V_o}{V_o + V_s}\sqrt{\lambda}.$$

Generalized DCM Duty Ratio as a Function of the Lambda Ratio

Substitution of Equation 5.51 into Equation 5.48 produces the expression for the generalized DCM duty ratio in terms of the λ-ratio.

$$D = \frac{V_2}{(V_2 - V_1)}\sqrt{\lambda}. \tag{5.52}$$

The generalized CCM duty ratio is readily recognized in Equation 5.52; the DCM duty ratio is thus related to the CCM duty ratio through the λ-ratio:

$$D = D_{CCM}\sqrt{\lambda}. \tag{5.53}$$

Equation 5.53 reveals that once the CCM duty ratio of a converter is known, the DCM duty ratio is also known without need of further analysis. Equation 5.53 yields the results of Example 5.7 and Exercise 5.7 directly.

EXAMPLE 5.8

A certain CCM converter has inductor voltages expressed by $V_1 = V_s - V_o$ and $V_2 = V_o$. Determine the CCM and DCM duty ratios.

Solution

Substitution of the inductor voltage expressions into Equation 5.4 yields the CCM duty ratio:

$$D_{CCM} = \frac{V_o}{2V_o - V_s}.$$

The DCM duty ratio follows directly from Equation 5.53.

$$D_{DCM} = \frac{V_o}{2V_o - V_s}\sqrt{\lambda}.$$

Conclusion

The generalized CCM and DCM duty ratio expressions are determined directly from the CCM inductor voltages without need of further analysis.

EXERCISE 5.8

A certain CCM converter has inductor voltages expressed by $V_1 = V_s$ and $V_2 = V_o$. Determine the CCM and DCM duty ratios.

Answers

$$D_{CCM} = \frac{V_o}{V_o - V_s}, \quad D_{DCM} = \frac{V_o}{V_o - V_s}\sqrt{\lambda}.$$

Generalized DCM Capacitance

As presented in Chapter 4, the capacitor current is equal to the difference between the inductor current and the load current. When the inductor current is zero, or when the inductor is disconnected from the load, the capacitor must sustain the load current. During the on-time, the DCM inductor current is expressed by Equation 5.5 with a minimum current of zero:

$$i_{L,1}(t) = \frac{V_1}{L}t, \; 0 \leq t \leq t_{on}. \tag{5.54}$$

During the off-time, the inductor current is expressed by Equation 5.6 until $t = t_x$:

$$i_{L,2}(t) = \frac{V_2}{L}(t - t_{on}) + I_{max}, \quad t_{on} \leq t \leq t_x. \tag{5.55}$$

The inductor current is zero from the extinction time to the end of the switching period:

$$i_{L,3}(t) = 0, \quad t_x \leq t \leq T. \tag{5.56}$$

If the load-connected duty ratio is unity, the capacitor current is the inductor current expressed by Equations 5.54 to 5.56 less the load current:

$$i_{C,1}(t) = \frac{V_1}{L} t - I_o, \quad 0 \leq t \leq t_{on}. \tag{5.57}$$

$$i_{C,2}(t) = \frac{V_2}{L}(t - t_{on}) + I_{max} - I_o, \quad t_{on} \leq t \leq t_x. \tag{5.58}$$

$$i_{C,3}(t) = -I_o, \quad t_x \leq t \leq T. \tag{5.59}$$

If the inductor is load-connected during the on-time, the capacitor must sustain the load current from $t = t_{on}$ for the remainder of the period. In particular, if $D_o = D$, the capacitor current is expressed by Equation 5.57 and

$$i_{C,3}(t) = -I_o, \quad t_{on} \leq t \leq T. \tag{5.60}$$

If the inductor is load-connected during the off-time, the capacitor must sustain the load current during the on-time and during the interval $t_x \leq t \leq T$. Consequently, if $D_o = 1 - D$, the capacitor current is expressed by Equation 5.58 and

$$i_{C,3}(t) = -I_o, \quad 0 \leq t \leq t_{on} \text{ and } t_x \leq t \leq T. \tag{5.61}$$

The DCM capacitor currents are plotted in Figure 5.10 along with their respective load-connected duty ratios.

The DCM capacitance is determined by application of the charge-balance principle to the waveforms illustrated in Figure 5.10. To compute the accumulated charge in the capacitor, the area above the time axis is formulated in terms of the λ-ratio for each load-connected duty ratio.

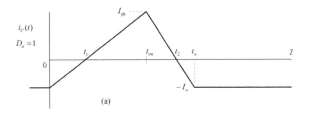

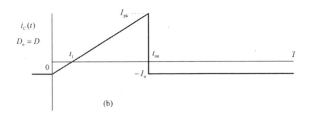

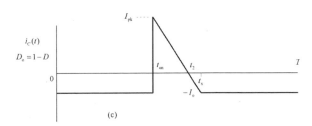

FIGURE 5.10   DCM capacitor currents and load-connected duty ratios.

$D_o = 1$

The accumulated charge is the area of the triangle with base $t_2 - t_1$ and height $I_{pk} = I_{max} - I_o$:

$$\Delta Q = \frac{1}{2}(t_2 - t_1)I_{pk}. \tag{5.62}$$

The peak capacitor current occurs at the same time as the maximum inductor current; $t = t_{on}$. In terms of D and f, the peak capacitor current from Equation 5.57 is

$$I_{pk} = \frac{DV_1}{Lf} - I_o. \tag{5.63}$$

With Equations 5.51 and 5.52, the DCM inductance is reformulated as

$$L = \frac{D_o D V_1}{2 I_o f} \sqrt{\lambda}. \tag{5.64}$$

With $D_o = 1$ in Equation 5.64, the peak capacitor current expressed by Equation 5.63 is reformulated in terms of the λ-ratio as

$$I_{pk} = \left(\frac{2}{\sqrt{\lambda}} - 1 \right) I_o. \tag{5.65}$$

The zero-crossing points in Figure 5.10a are the solutions of Equations 5.57 and 5.58 for t when $i_C(t) = 0$:

$$t_1 = \frac{L I_o}{V_1}. \tag{5.66}$$

$$t_2 = t_{on} + \frac{L}{V_2} \left(I_o - I_{max} \right). \tag{5.67}$$

With Equation 5.45, the base of the triangle is expressed as

$$t_2 - t_1 = \frac{D}{f} \left(1 - \frac{V_1}{V_2} \right) - \frac{L I_o \left(V_2 - V_1 \right)}{V_1 V_2}. \tag{5.68}$$

The first term on the RHS of Equation 5.68 is readily recognized as the extinction time expressed by Equation 5.44. With Equation 5.52, the extinction time is reformulated in terms of the λ-ratio as

$$t_x = \frac{\sqrt{\lambda}}{f}. \tag{5.69}$$

With Equations 5.51 and 5.69, the base of the triangle in terms of the λ-ratio is

$$t_2 - t_1 = \frac{2\sqrt{\lambda} - \lambda}{2f}. \tag{5.70}$$

Substitution of Equations 5.65 and 5.70 into Equation 5.62 yields the accumulated charge in the capacitor:

$$C\Delta V_{\mathrm{o}} = \frac{\left(2 - \sqrt{\lambda}\right)^2}{4f} I_{\mathrm{o}}. \tag{5.71}$$

From Equation 5.71 and the definition of ripple, the generalized DCM capacitance for a load-connected duty ratio of unity is

$$C = \frac{\left(2 - \sqrt{\lambda}\right)^2}{4Rfr}. \tag{5.72}$$

Equation 5.72 is the same capacitance expression developed for the DCM buck converter in Chapter 4.

$D_{\mathrm{o}} = D$

Substitution of $D_{\mathrm{o}} = D$ into Equation 5.64 yields the DCM inductance for a load-connected duty ratio equal to the on-time:

$$L = \frac{D^2 V_1}{2I_{\mathrm{o}}f} \sqrt{\lambda}. \tag{5.73}$$

With reference to Figure 5.10b, the accumulated charge in the capacitor is

$$\Delta Q = \frac{1}{2}\left(t_{\mathrm{on}} - t_1\right)\left(I_{\mathrm{max}} - I_{\mathrm{o}}\right). \tag{5.74}$$

With Equation 5.73, the zero-crossing time expressed by Equation 5.66 is reformulated in terms of the λ-ratio as

$$t_1 = \frac{D^2 \sqrt{\lambda}}{2f}. \tag{5.75}$$

Also with Equation 5.73, the peak capacitor current expressed by Equation 5.63 is formulated in terms of the λ-ratio as

$$I_{\mathrm{pk}} = \left(\frac{2}{D\sqrt{\lambda}} - 1\right) I_{\mathrm{o}}. \tag{5.76}$$

Substitution of Equations 5.75 and 5.76 into Equation 5.74 yields the accumulated charge:

$$CAV_o = \frac{\left(2-D\sqrt{\lambda}\right)^2}{4f\sqrt{\lambda}} I_o. \tag{5.77}$$

With the definition of ripple, the solution of Equation 5.77 for C yields the DCM capacitance for a load-connected duty ratio of D:

$$C = \frac{\left(2-D\sqrt{\lambda}\right)^2}{4Rfr\sqrt{\lambda}}. \tag{5.78}$$

$D_o = 1 - D$

Substitution of $D_o = 1 - D$ into Equation 5.64 yields a DCM inductance of

$$L = \frac{\left(1-D\right)DV_1}{2I_o f}\sqrt{\lambda}. \tag{5.79}$$

Substitution of Equation 5.79 into Equation 5.63 yields a peak capacitor current of

$$I_{pk} = \left[\frac{2}{\left(1-D\right)\sqrt{\lambda}} - 1\right] I_o. \tag{5.80}$$

In reference to Figure 5.10c, the accumulated charge is equal to the area of the triangle above the time-axis:

$$\Delta Q = \frac{1}{2}\left(t_2 - t_{on}\right)\left(I_{max} - I_o\right). \tag{5.81}$$

Subtraction of t_{on} from Equation 5.67 yields the base of the triangle as

$$t_2 - t_{on} = \frac{L}{V_2}\left(I_o - I_{max}\right). \tag{5.82}$$

Substitution of Equations 5.79 and 5.80 into Equation 5.82 yields the triangle base in terms of the λ-ratio:

$$t_2 - t_{on} = \frac{\left(1-D\right)\left(2-\sqrt{\lambda}\right)\left(\sqrt{\lambda}-D\right)}{2f}. \tag{5.83}$$

Substitution of Equations 5.80 and 5.83 into Equation 5.81 yields the accumulated charge:

$$CAV_o = \frac{\left(2-\sqrt{\lambda}\right)\left(\sqrt{\lambda}-D\right)}{4f\sqrt{\lambda}}\left[2-(1-D)\sqrt{\lambda}\right]I_o. \qquad (5.84)$$

The solution of Equation 5.84 for C is the DCM capacitance for a load-connected duty ratio of $D_o = 1 - D$:

$$C = \frac{\left(2-\sqrt{\lambda}\right)\left(\sqrt{\lambda}-D\right)}{4Rfr\sqrt{\lambda}}\left[2-(1-D)\sqrt{\lambda}\right]. \qquad (5.85)$$

The DCM Boost Converter

The duty ratio of the DCM boost converter is determined directly from Equations 5.35 and 5.53 as

$$D = \left(1 - \frac{V_s}{V_o}\right)\sqrt{\lambda}. \qquad (5.86)$$

From Equation 5.86, the output voltage is

$$V_o = \frac{V_s}{1 - \dfrac{D}{\sqrt{\lambda}}}, \qquad (5.87)$$

and the circuit transfer function is

$$\frac{V_o}{V_s} = \frac{1}{1 - \dfrac{D}{\sqrt{\lambda}}}. \qquad (5.88)$$

A comparison of Equations 5.36 and 5.88 reveals that the DCM boost converter is capable of larger magnitude output voltages than the CCM boost converter. Plots of the CCM and DCM converter transfer functions are shown in Figure 5.11.

In Figure 5.11 each transfer function is plotted for a duty ratio in the range of $0 \le D \le 0.8$. The DCM curves are plotted for λ-values in the range of $\lambda_{min} \le \lambda \le 0.9$, in which λ_{min} is 0.4, 0.2, 0.1, 0.05, and 0.01. The plots reveal that the smaller values of inductance result in larger values of output voltage for the same duty ratio.

The DCM capacitance for the boost converter is specified by Equation 5.85 because the inductor is load-connected during the power switch off-time. The next example illustrates the design of a DCM boost converter.

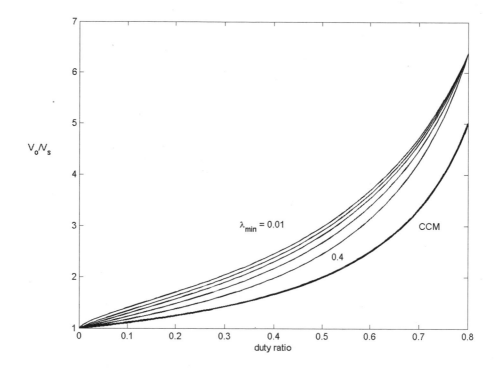

FIGURE 5.11   CCM and DCM transfer functions of the boost converter.

EXAMPLE 5.9

A boost converter is supplied by an automotive battery whose terminal voltage varies between 10 and 13.5 volts. The converter must output 120 VDC with 1% ripple to 144 Ω at a switching frequency of 25 kHz. Determine the DCM inductance and capacitance.

Solution

ON THE CD

The computations are performed by the program in Listing 5.1. The program is available in the Examples and Listings subfolder of Chapter 5 on the CD-ROM as file Example5_9.m.

LISTING 5.1   *L* and *C* Calculations for a DCM Boost Converter

```
close all, clear all, clc
Vs = 13.5;
Vo = 120;
```

```
R = 144;
r = 0.01;
Io = Vo/R;
f = 25E3;
Dccm = 1 - Vs/Vo
Lccm = R*Dccm/(2*f)*(1 - Dccm)^2
D = 0.5;
lda = D/Dccm;
L = lda*Lccm
slda = sqrt(lda);
C = (2 - slda)*(slda - D)*(2 - (1 - D)*slda)/(4*R*f*r*slda)
```

Conclusion

The larger source voltage results in the largest CCM inductance of 32 μH with a CCM duty ratio of 0.89. This unreasonably large duty ratio, however, is mitigated by the λ-ratio. A practical DCM duty ratio of 0.5 yields a λ-ratio of 0.56 and a DCM inductance of 18 μH. The DCM capacitance is 4.7 μF.

EXERCISE 5.9

Determine L and C for the converter in Example 5.9 if the computations are based on a source voltage of 10 V. If these component values are used in the circuit, what happens if the source voltage increases above 13.5 V?

Answers

$L = 10$ μH and $C = 4.6$ μF.

The DCM Buck/Boost Converter

With Equation 5.35 again and Equation 5.40, the duty ratio of the DCM buck/boost converter is expressed as

$$D = \frac{V_o}{V_o + V_s}\sqrt{\lambda}. \tag{5.89}$$

The solution of Equation 5.89 for V_o is the DCM output voltage expression

$$V_o = \frac{DV_s}{\sqrt{\lambda} - D}. \tag{5.90}$$

From Equation 5.90, the transfer function of the circuit is formulated as

$$\frac{V_o}{V_s} = \frac{D}{\sqrt{\lambda} - D}.$$

(5.91)

Equations 5.40 and 5.91 are plotted in Figure 5.12 for the same range of duty ratio and λ-values as in Figure 5.11.

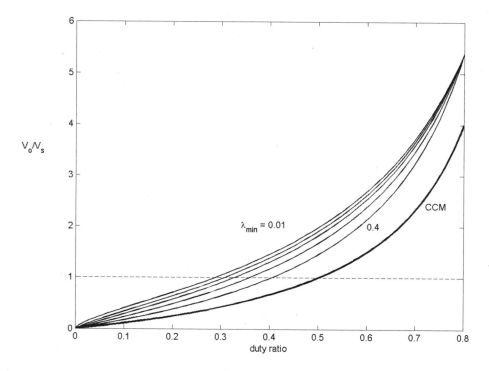

FIGURE 5.12   CCM and DCM transfer functions of the buck/boost converter.

As with the boost converter, the buck/boost converter also has larger magnitude output voltages for smaller inductance values for the same duty ratio. The boundary between the buck and boost modes for the CCM converter is always $D = \frac{1}{2}$. For the DCM converter, the boundary is dependent upon the circuit inductance.

The DCM capacitance for the buck/boost converter is also specified by Equation 5.85 because $D_o = 1 - D$. The next example illustrates the design of a DCM buck/boost converter.

EXAMPLE 5.10

A DCM buck/boost converter must output 36 VDC with 5% ripple from a source voltage that varies between 10 VDC and 54 VDC. The load current varies between 10 and 20 A. The PWM switching frequency is 20 kHz. Determine the DCM inductance and capacitance.

Solution

The solution is provided by the program in Listing 5.2 and is available as Example5_10.m in Chapter 5\Examples on the CD-ROM.

LISTING 5.2   *L* and *C* Calculations for a DCM Buck/Boost Converter

```
close all, clear all, clc
Vs = 10;
Vo = 36;
Io = 10;
R = Vo/Io;
r = 0.05;
f = 20E3;
Dccm = Vo/(Vo + Vs)
Lccm = R/(2*f)*(1 - Dccm)^2
D = 0.5;
lda = D/Dccm
L = lda*Lccm
slda = sqrt(lda);
C = (2 - slda)*(slda - D)*(2 - (1 - D)*slda)/(4*R*f*r*slda)
```

Conclusion

The low-end of the source voltage requires a large duty ratio (0.78) to boost the 10-V volt source to 36 VDC. The CCM inductance is computed with the largest load resistance and smallest operating duty ratio. The selected DCM duty ratio of ½ results in a λ-ratio of 0.64, a DCM inductance of 2.7 μH, and a DCM capacitance of 50 μF.

EXERCISE 5.10

Compute *L* and *C* for the converter in Example 5.10 if the CCM inductance is based on the largest output current. If these component values are used in the circuit, what happens if the load current decreases to 10 A?

Answers

$L = 1.4\ \mu H$ and $C = 100\ \mu F$.

H-BRIDGE DC-DC CONVERTERS

The buck, boost, and buck/boost converters have a common switching characteristic in regard to the storage inductance: one terminal of the inductor is fixed while the second terminal is switched between the source terminals or a source terminal and the R-C output network. The switching action is that of a single-pole, double-throw (SPDT) switch with one inductance terminal connected to the pole of the switch.

Other types of DC-DC converters result when both inductor terminals are connected to SPDT switches. Such a configuration is called an H-bridge because of the similarity of the circuit to the English letter.

The Watkins-Johnson Converter

The Watkins-Johnson converter has one inductor terminal that is alternately connected to the positive and negative source terminals. The other inductor terminal is alternately connected to the positive source terminal and the R-C output network. The functional schematic of the Watkins-Johnson converter is shown in Figure 5.13, in which the inductor connection is indicated for the first portion of the switching period.

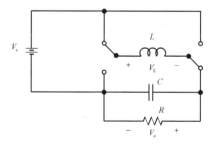

FIGURE 5.13   Functional schematic of the Watkins-Johnson converter.

Application of KVL to the circuit in Figure 5.13 yields the voltage across the inductor during the first part of the switching period:

$$V_{L,1}(t) = V_s - V_o. \tag{5.92}$$

Application of KVL to the circuit with the inductance connected alternately results in the inductor voltage during the second portion of the switching period:

$$V_{L,1}(t) = -V_s. \tag{5.93}$$

Substitution of Equations 5.92 and 5.93 into Equation 5.4 yields the duty ratio of the Watkins-Johnson converter:

$$D = \frac{V_s}{2V_s - V_o}. \tag{5.94}$$

The solution of Equation 5.94 for the ratio of the output voltage to the input voltage is the transfer function of the circuit:

$$\frac{V_o}{V_s} = \frac{2D - 1}{D}. \tag{5.95}$$

Equation 5.95 reveals that the output voltage of the Watkins-Johnson converter is negative with respect to the source voltage for duty ratios less than ½. Furthermore, the output voltage is of greater magnitude than the input voltage for $D < ½$. For duty ratios greater than ½, the output voltage is less than the input voltage but has the same polarity. The circuit thus performs buck/boost operations with the boost voltage of opposite polarity of the source. Equation 5.94 is plotted in Figure 5.14.

Since one terminal of the inductor in the Watkins-Johnson converter is always connected to the positive source terminal, the converter always operates in CCM.

The Inverse Watkins-Johnson Converter

The terminals of the inductor in the inverse Watkins-Johnson converter are alternately switched between the positive source terminal and the positive node of the R-C output network and between the negative source terminal and the positive node of the R-C network. The functional schematic of the inverse converter is shown in Figure 5.15.

Application of KVL to the circuit in Figure 5.15 with the inductor connected as indicated results in the same inductor voltage expression as Equation 5.92. Application of KVL with the inductor in the alternate connection results in the inductor voltage during the second portion of the switching period:

$$V_{L,2}(t) = V_o. \tag{5.96}$$

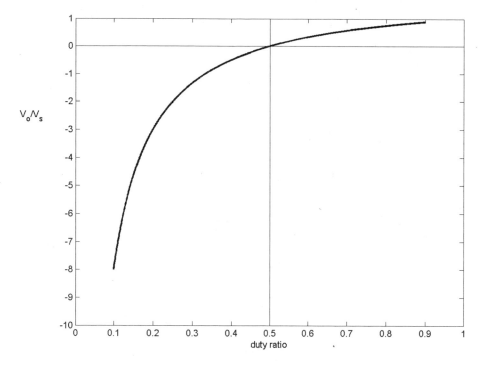

FIGURE 5.14 Transfer function of the Watkins-Johnson converter.

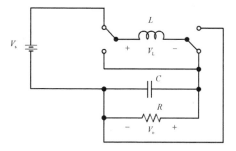

FIGURE 5.15 Functional schematic of the inverse Watkins-Johnson converter.

Substitution of Equations 5.92 and 5.96 into Equation 5.4 yields the duty ratio expression of the inverse converter:

$$D = \frac{V_o}{2V_o - V_s}. \tag{5.97}$$

The transfer function of the inverse converter is the solution of Equation 5.97 for V_o/V_s:

$$\frac{V_o}{V_s} = \frac{D}{2D-1}.$$

(5.98)

A plot of Equation 5.98 is shown in Figure 5.16.

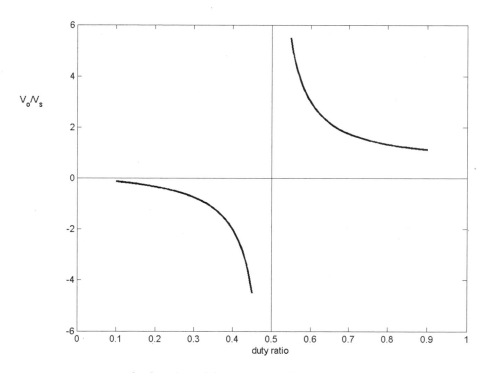

FIGURE 5.16   Transfer function of the inverse Watkins-Johnson converter.

As shown in Figure 5.16, the inverse converter is a boost-type converter with negative output voltages for $D < \frac{1}{2}$ and positive output voltages for $D > \frac{1}{2}$. The larger magnitude output voltages occur with a duty ratio near $\frac{1}{2}$.

Unlike the Watkins-Johnson converter, the inverse converter *is* capable of DCM operation. Directly from Equation 5.53, the DCM duty ratio is

$$D = \frac{V_o}{2V_o - V_s}\sqrt{\lambda}.$$

(5.99)

and the DCM transfer function is

$$\frac{V_o}{V_s} = \frac{D}{2D - \sqrt{\lambda}}.$$

(5.100)

The transfer function of the DCM inverse converter is plotted in Figure 5.17 for various values of λ-ratio.

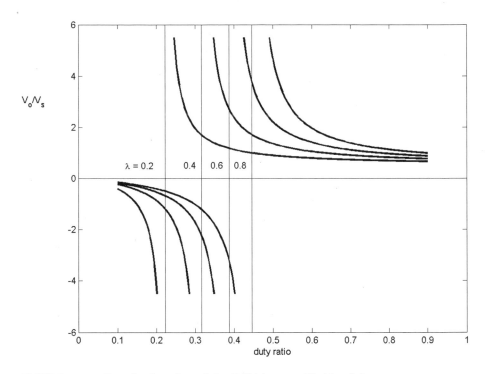

FIGURE 5.17 Transfer function of the DCM inverse Watkins-Johnson converter.

As Figure 5.17 indicates, the DCM inverse converter, as the boost converter, is capable of larger-magnitude output voltages than the equivalent CCM converter.

The Bridge Converter

The functional schematic of the bridge converter is shown in Figure 5.18.

Unlike the Watkins-Johnson converters, the inductor *and* the *R-C* load network are alternately switched across the terminals of the DC source.

In reference to Figure 5.18, application of KVL again results in the same expression as Equation 5.92. With the inductor and load network in the alternate connection, the inductor voltage is

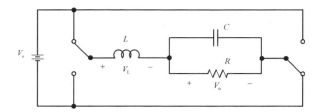

FIGURE 5.18   Functional schematic of the bridge converter.

$$V_{L,2}(t) = -\left(V_s + V_o\right).$$ (5.101)

Substitution of Equations 5.92 and 5.101 into Equation 5.4 yields the duty ratio expression

$$D = \frac{V_s + V_o}{2V_s},$$ (5.102)

and the transfer function expression

$$\frac{V_o}{V_s} = 2D - 1.$$ (5.103)

As the buck converter, the transfer function of the bridge converter is linear. The transfer function expressed by Equation 5.103 is plotted in Figure 5.19.

Figure 5.19 reveals that the bridge converter is indeed a buck-type converter and is capable of producing negative output voltages. The bridge converter always operates in CCM because the inductance is continually supplied by the source voltage.

The Current-Fed Bridge Converter

The final DC-DC converter considered in this chapter is the current-fed bridge converter. The functional schematic is shown in Figure 5.20.

In contrast to the bridge converter, the inductance in the current-fed converter is fixed, and the R-C network is alternately switched between an inductance terminal and the negative side of the source voltage.

Application of KVL with the R-C network connected as shown in Figure 5.20 once again yields the same expression as Equation 5.92. With the R-C network connected alternately, application of KVL yields

$$V_{L,2}(t) = V_s + V_o$$ (5.104)

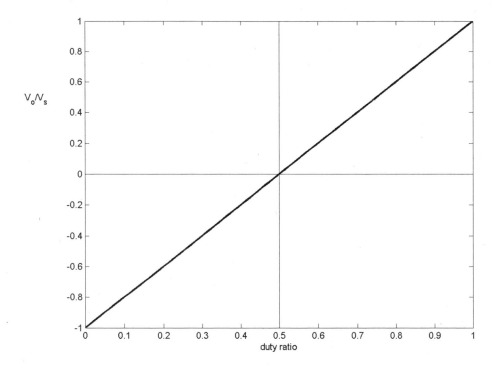

FIGURE 5.19 Transfer function of the bridge converter.

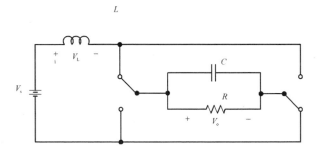

FIGURE 5.20 Functional schematic of the current-fed bridge converter.

Substitution of Equations 5.92 and 5.104 into Equation 5.4 yields the duty ratio of the current-fed bridge converter:

$$D = \frac{V_s + V_o}{2V_o}. \tag{5.105}$$

From Equation 5.105, the transfer function expression is derived as

$$\frac{V_o}{V_s} = \frac{1}{2D-1}.$$

(5.106)

Equation 5.106 is plotted in Figure 5.21

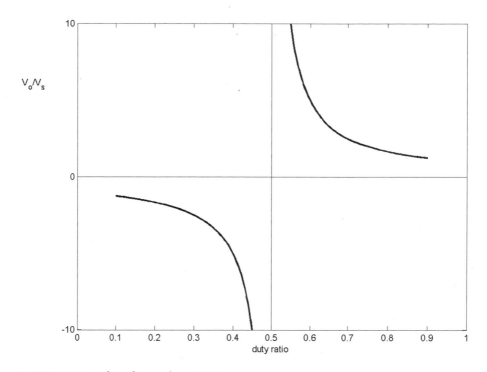

FIGURE 5.21   Plot of Equation 5.106.

Figure 5.21 reveals that the current-fed converter is a boost-type converter and is much like the inverse Watkins-Johnson converter but is capable of larger magnitude output voltages.

CHAPTER SUMMARY

The duty ratio derivation problem of any two-level converter has been effectively solved by the general theory of two-level converters. No longer subject to individual analysis, any converter whose storage inductance experiences two voltages that are constant over portions of the switching period and are of opposite polarity are subject to the theory.

The concept of the λ-ratio revolutionizes the theory of DCM converters. Quadratic equations are no longer necessary to expresses the output voltage of a two-level DCM converter. Once the expression for the CCM duty ratio is known, the DCM duty ratio and output voltage expressions directly and immediately follow. The λ-ratio and general theory effectively render the traditional methods of DC-DC converter analysis obsolete.

THE MATLAB TOOLBOX

ON THE CD

The functions listed below are located in the Toolbox subfolder of Chapter 5 on the CD-ROM.

LBOOST

Function *lboost.m* computes the minimum inductance required for a boost converter from the duty ratio, λ-ratio, load resistance, and PWM switching frequency.

Syntax

```
L = lboost(D, lambda, R, f)
```

CBOOST

Function *cboost.m* computes the minimum required capacitance for a boost converter from the duty ratio, ripple, λ-ratio, load resistance, and PWM switching frequency.

Syntax

```
C = cboost(D, r, lambda, R, f)
```

LBKBST

Function *lbkbst.m* computes the minimum inductance required for a buck/boost converter from the duty ratio, λ-ratio, load resistance, and PWM switching frequency.

Syntax

```
L = lbkbst(D, lambda, R, f)
```

CBKBST

Function *cbkbst.m* computes the minimum required capacitance for a buck/boost converter from the duty ratio, ripple, λ-ratio, load resistance, and PWM switching frequency.

Syntax

```
C = lbkbst(D, r, lambda, R, f)
```

PROBLEMS

For the design problems, compute the minimum inductance and capacitance that meets the circuit specification.

Problem 1

Design a CCM boost converter to output 48 VDC at 10 W with 2% ripple from a source that varies between 10 and 14 VDC with a PWM switching frequency of 25 kHz.

Problem 2

Design a DCM boost converter to output 48 VDC at 10 W with 2% ripple from a source that varies between 10 and 14 VDC with a PWM switching frequency of 25 kHz.

Problem 3

Design a CCM boost converter to output 90 VDC at 0.5 to 1 A with 5% ripple from a source that varies between 30 and 42 VDC with a PWM switching frequency of 20 kHz.

Problem 4

Design a DCM boost converter to output 90 VDC at 0.5 to 1 A with 5% ripple from a source that varies between 30 and 42 VDC with a PWM switching frequency of 20 kHz.

Problem 5

Design a CCM buck/boost converter to output 12 VDC at 50 W with 1% ripple from a source that varies between 10 and 14 VDC with a PWM switching frequency of 25 kHz.

Problem 6

Design a DCM buck/boost converter to output 12 VDC at 50 W with 1% ripple from a source that varies between 10 and 14 VDC with a PWM switching frequency of 25 kHz.

Problem 7

Design a CCM buck/boost converter to output 36 VDC at 2 to 4 A with 3% ripple from a source that varies between 30 and 42 VDC with a PWM switching frequency of 20 kHz.

Problem 8

Design a DCM buck/boost converter to output 36 VDC at 2 to 4 A with 3% ripple from a source that varies between 30 and 42 VDC with a PWM switching frequency of 20 kHz.

Problem 9

Derive the transistor and diode dissipation expressions and the efficiency expression for a practical boost converter implemented with a BJT and diode.

Problem 10

Derive the transistor and diode dissipation expressions and the efficiency expression for a practical buck/boost converter implemented with a BJT and diode.

Problem 11

Derive the CCM inductance expression for the inverse Watkins-Johnson converter.

Problem 12

Derive the CCM capacitance expression for the inverse Watkins-Johnson converter.

6 Dynamic Modeling and Simulation of DC-DC Converters

In This Chapter

- Introduction
- State-Space Modeling of Linear Systems
- State-Space Modeling of Linear Circuits
- Simulation of State-Space Models in MATLAB
- State-Space Modeling of DC-DC Converters
- Simulation of DC-DC Converters in MATLAB
- Chapter Summary
- The MATLAB Toolbox
- Problems

INTRODUCTION

The prior chapters on DC-DC converters presented circuit models without dynamic characteristics. Those models present no indication of how the converter behaves under transient conditions. The neglect of the dynamic behavior, however, provided an elegant and highly useful body of theory that provides the circuit designer with a valuable commodity: a starting place.

Once the inductance and capacitance values are determined from the CCM or DCM equations, those circuit parameters are used in simulation to observe the transient behavior of the circuit and to understand the effects of an increase or decrease in inductance and/or capacitance.

The mathematical transient features of a circuit must be expressed by differential equations. To facilitate the simulation process, the differential equations are formulated into the *state-space* model of the circuit.

STATE-SPACE MODELING OF LINEAR SYSTEMS

The state-space model of a linear system is a set of simultaneous, or *coupled*, first-order differential equations. An example of two coupled, first-order differential equations is the equation set

$$\frac{dx_1}{dt} = a_{11}x_1 + a_{12}x_2 + f(t)$$
$$\frac{dx_2}{dt} = a_{21}x_1 + a_{22}x_2, \tag{6.1}$$

in which x_1 and x_2 are the *state variables*, $a_{11} \ldots a_{22}$ are constant coefficients, and $f(t)$ is a forcing function. The state-space model is a reformulation of the coupled equations into the matrix-vector form

$$\dot{x} = Ax + Bu. \tag{6.2}$$

In the case of Equation 6.1, vectors **x** and **u** in the state-space model are

$$\mathbf{x} = \begin{bmatrix} x_1 \\ x_2 \end{bmatrix}, \quad \mathbf{u} = f(t),$$

and matrices **A** and **B** are

$$A = \begin{bmatrix} a_{11} & a_{12} \\ a_{21} & a_{22} \end{bmatrix}, \quad \mathbf{B} = \begin{bmatrix} 1 \\ 0 \end{bmatrix}.$$

In general, for n first-order linear differential equations with m forcing functions, the state-space model is expressed by Equation 6.2 with

$$A = \begin{bmatrix} a_{11} & a_{12} & \cdots & a_{1n} \\ a_{21} & & & \vdots \\ \vdots & & & \vdots \\ a_{n1} & a_{n2} & \cdots & a_{nn} \end{bmatrix}, \quad \mathbf{x} = \begin{bmatrix} x_1 \\ x_2 \\ \vdots \\ x_n \end{bmatrix}, \quad B = \begin{bmatrix} b_{11} & b_{12} & \cdots & b_{1m} \\ b_{21} & & & \vdots \\ \vdots & & & \vdots \\ b_{n1} & b_{n2} & \cdots & b_{nm} \end{bmatrix}, \quad \mathbf{u} = \begin{bmatrix} f_1(t) \\ f_2(t) \\ \vdots \\ f_m(t) \end{bmatrix}. \tag{6.3}$$

In Equation 6.3, **A** is the *state transition matrix* with dimensions $[n \times n]$, **x** is the *state vector* with dimensions $[n \times 1]$, **B** is the *control input matrix* with dimensions $[n \times m]$, and **u** is the *control vector* with dimensions $[m \times 1]$. An nth-order linear differential equation is readily formed into the state-space model by the definition of n state variables. The next example illustrates the method.

EXAMPLE 6.1

Find the state-space model of the second-order differential equation

$$\ddot{x} + a\dot{x} + bx = f(t). \tag{6.4}$$

Solution

First, the dependent variable x in Equation 6.4 is renamed state variable x_1:

$$\ddot{x}_1 + a\dot{x}_1 + bx_1 = f(t). \tag{6.5}$$

Next, state variable x_1 is differentiated with respect to time and assigned to the state variable x_2:

$$\dot{x}_1 = x_2. \tag{6.6}$$

Substitution of x_2 for the first derivative of x_1 in Equation 6.5 results in

$$\ddot{x}_1 + ax_2 + bx_1 = f(t). \tag{6.7}$$

From Equation 6.6, the first derivative of x_2 is also the second derivative of x_1:

$$\dot{x}_2 = \ddot{x}_1. \tag{6.8}$$

Substitution of Equation 6.8 into Equation 6.7 results in

$$\dot{x}_2 + ax_2 + bx_1 = f(t). \tag{6.9}$$

The solution of Equation 6.9 for the first derivative of x_2 is

$$\dot{x}_2 = -bx_1 - ax_2 + f(t). \tag{6.10}$$

Equations 6.6 and 6.10 form the coupled equations of the state-space model:

$$\begin{bmatrix} \dot{x}_1 \\ \dot{x}_2 \end{bmatrix} = \begin{bmatrix} 0 & 1 \\ -b & -a \end{bmatrix} \begin{bmatrix} x_1 \\ x_2 \end{bmatrix} + \begin{bmatrix} 0 \\ 1 \end{bmatrix} f(t). \tag{6.11}$$

Conclusion

The second-order differential equation is transformed into two first-order differential equations. The first derivative of the first state variable becomes itself a state variable in the state-space model. In general, the state-space method generates n first-order differential equations from an nth-order differential equation with one forcing function:

$$\dot{x}_1 = x_2$$
$$\dot{x}_2 = x_3$$
$$\vdots$$
$$\dot{x}_{n-1} = x_{n-2}. \qquad (6.12)$$

The nth first-order equation is a linear combination of the state variables and the forcing function:

$$\dot{x}_n = -a_1 x_1 - a_2 x_2 - \ldots - a_{n-1} x_{n-1} + f(t). \qquad (6.13)$$

EXERCISE 6.1

Find the state-space model of the third-order differential equation

$$\dddot{x} + a\ddot{x} + b\dot{x} + cx = f(t).$$

Answer

$$\begin{bmatrix} \dot{x}_1 \\ \dot{x}_2 \\ \dot{x}_3 \end{bmatrix} = \begin{bmatrix} 0 & 1 & 0 \\ 0 & 0 & 1 \\ -c & -b & -a \end{bmatrix} \begin{bmatrix} x_1 \\ x_2 \\ x_3 \end{bmatrix} + \begin{bmatrix} 0 \\ 0 \\ 1 \end{bmatrix} f(t).$$

STATE-SPACE MODELING OF LINEAR CIRCUITS

The state variables of a linear circuit are voltages and currents. The state-space model is greatly facilitated if the state variables are chosen as inductor current and capacitor voltage because of the first-derivative circuit laws:

$$V_L = L \frac{di}{dt} \qquad (6.14)$$

and

$$i_C = C\frac{dv}{dt}. \tag{6.15}$$

In general, the number of inductors plus the number of capacitors determines the number of state variables and, hence, the order of the state-space system. The sources in the circuit are the forcing functions and comprise the control vector. In general, the number of sources m determines the dimensions of the control vector and the control input matrix. The next example illustrates how to determine the state-space model after application of KVL and KCL.

EXAMPLE 6.2

Develop the state-space model of the circuit shown in Figure 6.1.

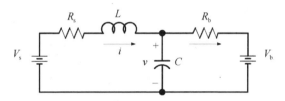

FIGURE 6.1   Circuit for Example 6.2.

Solution

The state variables are chosen as inductor current i and capacitor voltage v. Application of KVL around the left circuit loop yields the voltage equation

$$V_s = iR_s + L\frac{di}{dt} + v. \tag{6.16}$$

Application of KCL at the common node of the inductor and capacitor yields the current equation

$$i = C\frac{dv}{dt} + \frac{v - V_b}{R_b}. \tag{6.17}$$

A rearrangement of Equations 6.16 and 6.17 with the first derivatives on the lefthand side (LHS) yields

$$\frac{di}{dt} = -\frac{R_s}{L}i - \frac{1}{L}v + \frac{1}{L}V_s \qquad (6.18)$$

and

$$\frac{dv}{dt} = \frac{1}{C}i - \frac{1}{R_bC}v + \frac{V_b}{R_bC}. \qquad (6.19)$$

From Equations 6.18 and 6.19, the state-space model emerges as

$$\frac{d}{dt}\begin{bmatrix} i \\ v \end{bmatrix} = \begin{bmatrix} -\dfrac{R_s}{L} & -\dfrac{1}{L} \\ \dfrac{1}{C} & -\dfrac{1}{R_bC} \end{bmatrix}\begin{bmatrix} i \\ v \end{bmatrix} + \begin{bmatrix} \dfrac{1}{L} & 0 \\ 0 & \dfrac{1}{R_bC} \end{bmatrix}\begin{bmatrix} V_s \\ V_b \end{bmatrix}. \qquad (6.20)$$

Conclusion

The coupled equations are readily obtained by application of KVL around loops that contain inductors and application of KCL at capacitor-connected nodes.

EXERCISE 6.2

Develop the state-space model of the circuit shown in Figure 6.2

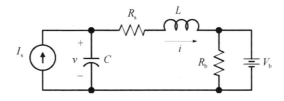

FIGURE 6.2   Circuit for Exercise 6.2.

Answer

$$\frac{d}{dt}\begin{bmatrix} i \\ v \end{bmatrix} = \begin{bmatrix} -\dfrac{R_s}{L} & \dfrac{1}{L} \\ -\dfrac{1}{C} & 0 \end{bmatrix}\begin{bmatrix} i \\ v \end{bmatrix} + \begin{bmatrix} -\dfrac{1}{L} & 0 \\ 0 & \dfrac{1}{C} \end{bmatrix}\begin{bmatrix} I_s \\ V_b \end{bmatrix}.$$

SIMULATION OF STATE-SPACE MODELS IN MATLAB

MATLAB is an excellent platform for state-space model simulation because of the matrix data structure upon which it is based. MATLAB also has function files that compute numerical solutions to coupled differential equations. Two m-files are required to simulate a state-space model. The first file contains the circuit parameters along with the definitions of the state transition and control input matrices. The second file contains the control function and the state derivative function described by Equation 6.2. The following examples illustrate how to simulate a state-space model in MATLAB.

EXAMPLE 6.3

Compute and plot the inductor current response of the circuit in Example 6.1 with circuit parameters $V_s = 170$, $R_s = 0.3$ Ω, $L = 10$ mH, $C = 220$ uF, $R_b = 0.5$ Ω, $V_b = 144$ VDC, and zero initial conditions.

Solution

ON THE CD

The solution code is provided in Listings 6.1 and 6.2. The solution programs are available on the CD-ROM as Example6_3.m and ckt6_3.m in the Examples and Listing subfolder of Chapter 6.

LISTING 6.1   MATLAB Program to Compute and Plot the Inductor Current Response

```
close all, clear all, clc
global A B u
mH = 1E-3;
uF = 1E-6;
Vs = 170;
Vb = 144;
Rs = 0.3;
Rb = 0.5;
L = 10*mH;
C = 220*uF;
RC = Rb*C;
A = [-Rs/L -1/L; 1/C -1/RC];
B = [1/L 0; 0 1/RC];
u = [Vs; Vb];
X0 = [0; 0];
[t, X] = ode23('ckt6_3', [0 0.1], X0);
IL = X(:, 1);
plot(t, IL), grid
```

LISTING 6.2 MATLAB Function File to Compute the State Derivative

```
function dx = ckt6_3(t, x)
global A B u
dx = A*x + B*u;
```

Conclusion

The program in Listing 6.1 makes use of the *global* statement. This command allows the variables in the list to be available to and used by any function that includes the same *global* statement.

Once the variables, matrices, vectors, and initial conditions are defined, the ordinary differential equation solver *ode23.m* is invoked with a filename, a time vector, and an initial condition vector. The file processed by *ode23* is *ckt6_3.m*, shown in Listing 6.2, the file that contains the coded form of Equation 6.2.

The *ode23* function applies the *Runge-Kutta* algorithm to solve a system of differential equations. The system of equations must appear in the file with the name that appears in single quotes in the *ode23* function call. The algorithm computes a time vector **t** and a state matrix **X**. The time vector contains elements that begin and end with the time values specified in the function call. Matrix **X** contains the values of the state variables. The first state variable is placed in the first column of **X**, the second state variable is placed in the second column, and so on. MATLAB statement $IL = X(:, 1)$ assigns all the elements in the first column of **X** to the variable *IL*. The colon operator $(:)$ is synonymous with the word *all* as pertains to rows or columns of a matrix. Therefore, vector **IL** is assigned the values of the elements of *all* rows of the first column of **X**. The inductor current is plotted in Figure 6.3.

As indicated in Figure 6.3, the inductor current reaches the steady-state value of

$$I_{ss} = \frac{V_s - V_b}{R_s + R_b} = \frac{170 - 144}{0.3 + 0.5} = 32.5 \text{ A.}$$

EXERCISE 6.3

Simulate and plot the capacitor voltage of the circuit in Exercise 6.2 with circuit parameters $C = 10 \text{ μF}$, $L = 5 \text{ mH}$, $I_s = 10 \text{ A}$, $R_s = 0.3 \text{ Ω}$, and $R_b = 100 \text{ Ω}$, within the time interval 0 to 0.1 seconds.

Answer

See Figure 6.4.

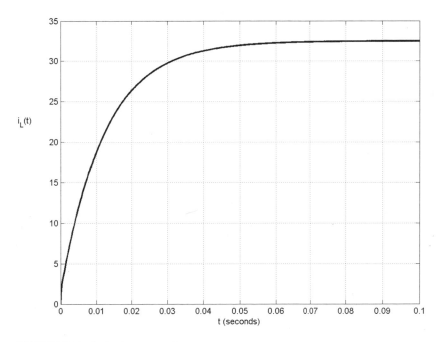

FIGURE 6.3   Inductor current response.

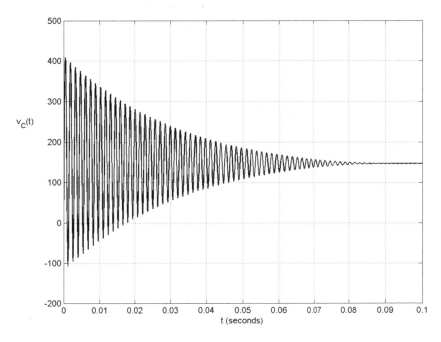

FIGURE 6.4   Solution to Exercise 6.3.

The forcing functions of the circuits in Example 6.3 and Exercise 6.3 are constants and as such are defined in the *global* statement. Time-varying sources must be computed in the function file along with the state derivative. The next example shows how to simulate circuits with time-varying sources.

EXAMPLE 6.4

Simulate the circuit in Example 6.2 with $V_s = 50 \sin(120\pi t) + 170$. Plot the inductor current and the capacitor voltage.

Solution

The solution code is provided in Listings 6.3 and 6.4. The programs are available on the CD-ROM as Example6_4.m and ckt6_4.m in folder Chapter 6\Examples and Listings.

LISTING 6.3 Circuit Simulation with a Time-Varying Source

```
close all, clear all, clc
global A B Vb
Vb = 144;
mH = 1E-3;
uF = 1E-6;
Rs = 0.3;
Rb = 0.5;
L = 10*mH;
C = 220*uF;
RC = Rb*C;
A = [-Rs/L -1/L; 1/C -1/RC];
B = [1/L 0; 0 1/RC];
X0 = [0; 0];
[t, X] = ode23('ckt6_4', [0 0.1], X0);
IL = X(:, 1);
VC = X(:, 2);
subplot(1,2,1), plot(t, IL, 'k'), grid
axis square, title('i_L(t)'), xlabel('t (sec)')
subplot(1,2,2), plot(t, VC, 'k'), grid
axis square, title('v_C(t)'), xlabel('t (sec)')
```

LISTING 6.4 Function File for Example 6.4

```
function dx = ckt6_4(t, x)
global A B Vb
Vs = 50*sin(120*pi*t) + 170;
u = [Vs; Vb];
dx = A*x + B*u;
```

Conclusion

The circuit state variables are plotted in Figure 6.5.

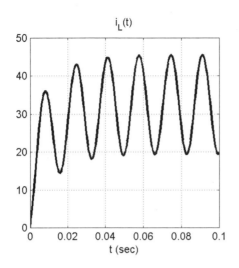
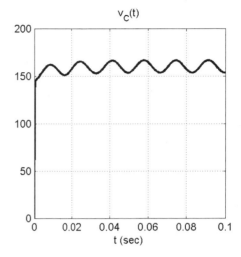

FIGURE 6.5   State variables for Example 6.4.

The time-varying source is readily accommodated in the function file. The time variable t is computed by *ode23* and passed to the function file. The subplot command is used to divide the screen into two graphs. The general syntax of the command is *subplot(m,n,p)*, in which m and n are the numbers of rows and columns, respectively, into which the screen is divided, and p is a particular graph number. Thus, *subplot(1,2,2)*, for example, accesses graph 2 (p) in row 1 (m), column 2 (n).

The axis command has several features. In this example, the *square* parameter is used to make the vertical and horizontal axes of equal length. The *title* command provides a plot caption, while *xlabel* generates a horizontal axis label. The underscore is used in the text strings of *title* to create a subscript.

EXERCISE 6.4

Simulate the circuit in Exercise 6.3 with the time-varying current source $I_s = 10\text{abs}\left[\sin\left(40\pi t\right)\right]$, in which *abs* is the absolute value. Plot the state variables.

Answer

See Figure 6.6.

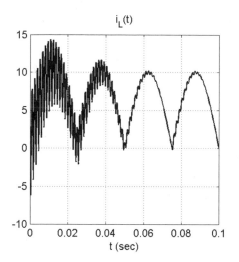
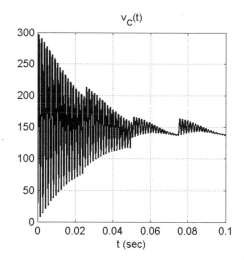

FIGURE 6.6   Solution to Exercise 6.4.

STATE-SPACE MODELING OF DC-DC CONVERTERS

The DC-DC converter has the special distinction of being a system of *variable structure*; that is, the circuit topology changes in accordance with the switching action of the semiconductor devices. The state-space model, therefore, must describe the dynamic behavior of the circuit for each portion of the switching cycle. In the case of the buck converter, the first structure contains the source voltage V_s; the second structure is source free.

State-Space Model of the Buck Converter

The two circuit topologies of the buck converter are illustrated in Figure 6.7.

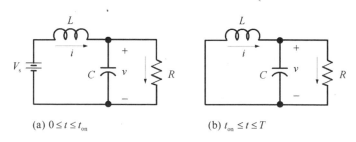

FIGURE 6.7   Topologies of the buck converter.

With reference to Figure 6.7a, KVL and KCL yield the differential equations that describe the dynamic behavior of the buck converter during the on-time:

$$V_s = L\frac{di}{dt} + v \tag{6.21}$$

and

$$i = C\frac{dv}{dt} + \frac{v}{R}. \tag{6.22}$$

During the off-time, the inductor is disconnected from the source and short-circuited to ground. In reference to Figure 6.7b, Equation 6.21 becomes

$$0 = L\frac{di}{dt} + v, \tag{6.23}$$

while Equation 6.22 remains unchanged. Equations 6.21 and 6.23 are combined by the definition of a binary control switch u:

$$u = \begin{cases} 1, & 0 \leq t \leq t_{on} \\ 0, & t_{on} \leq t \leq T \end{cases}. \tag{6.24}$$

With Equation 6.24, Equations 6.21 and 6.23 are combined to form

$$uV_s = L\frac{di}{dt} + v. \tag{6.25}$$

The solutions of Equations 6.22 and 6.25 for the derivatives yield the dynamic equations of the buck converter over the entire switching period:

$$\frac{di}{dt} = -\frac{v}{L} + \frac{V_s}{L}u. \tag{6.26}$$

$$\frac{dv}{dt} = \frac{1}{C}i - \frac{1}{RC}v. \tag{6.27}$$

With Equations 6.26 and 6.27, the variable structure state-space model of the buck converter is formulated as

$$\frac{d}{dt}\begin{bmatrix} i \\ v \end{bmatrix} = \begin{bmatrix} 0 & -\dfrac{1}{L} \\ \dfrac{1}{C} & -\dfrac{1}{RC} \end{bmatrix}\begin{bmatrix} i \\ v \end{bmatrix} + \begin{bmatrix} \dfrac{V_s}{L} \\ 0 \end{bmatrix} u. \qquad (6.28)$$

The buck converter is thus a single-input linear system with constant coefficients.

State-Space Model of the Boost Converter

The circuit topologies of the boost converter are illustrated in Figure 6.8.

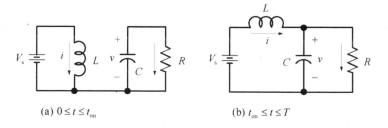

(a) $0 \le t \le t_{on}$ (b) $t_{on} \le t \le T$

FIGURE 6.8 Topologies of the boost converter.

With reference to Figure 6.8a, application of KVL and KCL yields, respectively,

$$V_s = L\frac{di}{dt} \qquad (6.29)$$

and

$$0 = C\frac{dv}{dt} + \frac{v}{R}. \qquad (6.30)$$

Equations 6.29 and 6.30 describe the dynamics of the boost converter during the on-time. When the primary switch is off, the differential equations of the boost converter are the same as the on-time equations of the buck converter: Equations 6.21 and 6.22. With the binary control described by Equation 6.24, Equations 6.21, and 6.29 are combined in the form of

$$V_s = L\frac{di}{dt} + (1-u)v. \qquad (6.31)$$

Similarly, Equations 6.22 and 6.30 are combined to form

$$(1-u)i = C\frac{dv}{dt} + \frac{v}{R}. \tag{6.32}$$

The solutions of Equations 6.31 and 6.32 for the derivatives yield the dynamic equations of the boost converter throughout the switching period:

$$\frac{di}{dt} = -\frac{1}{L}v + \frac{1}{L}V_s + \frac{1}{L}vu. \tag{6.33}$$

$$\frac{dv}{dt} = \frac{1}{C}i - \frac{1}{RC}v - \frac{1}{C}iu. \tag{6.34}$$

With the source voltage and binary switch treated as separate control inputs, the variable structure state-space model of the boost converter is formulated as

$$\frac{d}{dt}\begin{bmatrix} i \\ v \end{bmatrix} = \begin{bmatrix} 0 & -\dfrac{1}{L} \\ \dfrac{1}{C} & -\dfrac{1}{RC} \end{bmatrix}\begin{bmatrix} i \\ v \end{bmatrix} + \begin{bmatrix} \dfrac{v}{L} & \dfrac{1}{L} \\ -\dfrac{i}{C} & 0 \end{bmatrix}\begin{bmatrix} u \\ V_s \end{bmatrix}. \tag{6.35}$$

Unlike the buck converter, the state-space model of the boost converter has two forcing functions. Additionally, the control input matrix does not have constant elements; it is a function of the state variables.

State-Space Model of the Buck/Boost Converter

The circuit topologies of the buck/boost converter are shown in Figure 6.9.

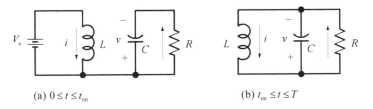

(a) $0 \leq t \leq t_{on}$ (b) $t_{on} \leq t \leq T$

FIGURE 6.9   Topologies of the buck/boost converter.

Application of KVL and KCL to the circuit in Figure 6.9a results in the same equations as the boost converter during the on-time: Equations 6.29 and 6.30. Application of KCL and KVL to the circuit in Figure 6.9b yields

$$0 = L\frac{di}{dt} + v \tag{6.36}$$

and

$$i = C\frac{dv}{dt} + \frac{v}{R}. \tag{6.37}$$

The binary control variable allows Equations 6.29 and 6.36 to be combined into

$$uV_s = L\frac{di}{dt} + (1-u)v, \tag{6.38}$$

while Equations 6.30 and 6.37 are combined as

$$(1-u)i = C\frac{dv}{dt} + \frac{v}{R}. \tag{6.39}$$

The solutions of Equations 6.38 and 6.39 for the derivatives are

$$\frac{di}{dt} = -\frac{1}{L}v + \frac{1}{L}(v+V_s)u \tag{6.40}$$

and

$$\frac{dv}{dt} = \frac{1}{C}i - \frac{1}{RC}v - \frac{1}{C}iu. \tag{6.41}$$

With Equations 6.40 and 6.41, the variable structure state-space model of the buck/boost converter emerges as

$$\frac{d}{dt}\begin{bmatrix} i \\ v \end{bmatrix} = \begin{bmatrix} 0 & -\dfrac{1}{L} \\ \dfrac{1}{C} & -\dfrac{1}{RC} \end{bmatrix}\begin{bmatrix} i \\ v \end{bmatrix} + \begin{bmatrix} \dfrac{1}{L}(v+V_s) \\ -\dfrac{i}{C} \end{bmatrix}u. \tag{6.42}$$

As the state-space model of the buck converter, the buck/boost model is a single-input system. As the state-space model of the boost converter, the control input matrix of the buck/boost model is a function of the state variables.

SIMULATION OF DC-DC CONVERTERS IN MATLAB

The Runge-Kutta algorithm employed in *ode23* uses a variable step length to improve computational accuracy. The step-length is decreased for large-magnitude derivatives and increased for smaller magnitude derivatives. Unfortunately, the values of physical components typically result in large-magnitude derivatives and long simulation times. However, this issue is readily addressed if the differential equations and state-space models are normalized with respect to the switching frequency. The normalization process alleviates the requirement to compute large derivatives and significantly decreases the simulation time.

Normalization with Respect to Frequency

Since the units of time are seconds and the units of frequency are the inverse of seconds, a normalization results from the product of frequency and time:

$$\tau = ft. \tag{6.43}$$

The solution of Equation 6.43 for time is

$$t = \frac{1}{f}\tau. \tag{6.44}$$

Differentiation of Equation 6.44 results in

$$dt = \frac{1}{f}d\tau. \tag{6.45}$$

A differential equation is normalized upon substitution of Equations 6.44 and 6.45 for t and dt, respectively. The normalization process is illustrated by the following examples and exercises.

EXAMPLE 6.5

Normalize the differential equations and the subsequent state-space model from Example 6.1 with respect to a frequency of 10 Hz. The system parameters are $a = 10$ and $b = 10000$; the forcing function is $f(t) = 100\cos(20\pi t)$.

Solution

Expressed in terms of the given system parameters and the forcing function, Equations 6.6 and 6.10 become

$$\frac{dx_1}{dt} = x_2 \tag{6.46}$$

and

$$\frac{dx_2}{dt} = -10000x_1 - 10x_2 + 100\cos(20\pi t). \tag{6.47}$$

Substitution of Equations 6.44 and 6.45 into Equations 6.46 and 6.47 with $f = 10$ results in the normalized equations

$$\frac{dx_1}{d\tau} = \frac{1}{10}x_2 \tag{6.48}$$

and

$$\frac{dx_2}{d\tau} = -1000x_1 - x_2 + 10\cos(2\pi\tau). \tag{6.49}$$

With Equations 6.48 and 6.49, the normalized state-space model is

$$\frac{d}{d\tau}\begin{bmatrix} x_1 \\ x_2 \end{bmatrix} = \begin{bmatrix} 0 & \frac{1}{10} \\ -1000 & -1 \end{bmatrix}\begin{bmatrix} x_1 \\ x_2 \end{bmatrix} + \begin{bmatrix} 0 \\ 10 \end{bmatrix}\cos(2\pi\tau). \tag{6.50}$$

Conclusion

Equations 6.48 and 6.49 reveal that the normalization process has reduced the magnitudes of the derivatives. Since the independent variable τ is dimensionless, the period of the forcing function has been normalized to 2π radians.

EXERCISE 6.5

Normalize the differential equation

$$\ddot{x} + 2000\dot{x} + 1000x = 1000\cos(2000\pi t) + 2000\sin(4000\pi t)$$

with respect to an appropriate frequency.

Answer

$$\frac{dx_1}{d\tau} = \frac{1}{1000}x_2$$

$$\frac{dx_2}{d\tau} = -x_1 - 2x_2 + \cos(2\pi\tau) + 2\sin(4\pi\tau).$$

EXAMPLE 6.6

Simulate the system described by Equations 6.46 and 6.47 from 0 to 1 second with zero initial conditions.

Solution

ON THE CD

The solution code is provided in Listings 6.5 and 6.6. The programs are located in folder Chapter 6\Examples and Listings on the CD-ROM as files Example6_6.m and ckt6_6.m.

LISTING 6.5   Simulation of Equations 6.46 and 6.47

```
close all, clear all, clc
global A B
a = 10;
b = 10000;
A = [0 1; -b -a];
B = [0; 1];
X0 = [0; 0];
tic
[t, X] = ode23('ckt6_6', [0 1], X0);
toc
x1 = X(:, 1);
x2 = X(:, 2);
subplot(1,2,1), plot(t, x1), grid
axis square
title('x_1'), xlabel('t (sec)')
subplot(1,2,2), plot(t, x2), grid
axis square
title('x_2'), xlabel('t (sec)')
```

LISTING 6.6   Function File for Example 6.6

```
function dx = ckt6_6(t, x)
global A B
u = 100*cos(20*pi*t);
dx = A*x + B*u;
```

Conclusion

The results of the simulation are plotted in Figure 6.10.

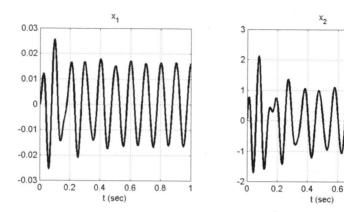

FIGURE 6.10 Simulation results of Example 6.6.

The code in Listing 6.5 makes use of *tic* and *toc*, which display the elapsed time between the executions of the commands.

EXERCISE 6.6

Simulate the circuit in Example 6.6 from 0 to 2 seconds and determine the number of computed time points with the *length* command. Compare this number with the number of points from the 0 to 1 second simulation.

Answers

[0 1], 327 points
[0 2], 625 points.

EXAMPLE 6.7

Simulate the normalized system described by Equation 6.50 and verify that the results are identical to Figure 6.10.

Solution

ON THE CD

The solution code is provided in Listings 6.7 and 6.8. The code is contained in files Example6_7.m and ckt6_7.m on the CD-ROM in the Examples and Listings subfolder of Chapter 6.

LISTING 6.7   Simulation of Equation 6.50

```
close all, clear all, clc
global A B
A = [0 0.1; -1000 -1];
B = [0; 10];
X0 = [0; 0];
tic
[tau, X] = ode23('ckt6_7', [0 10], X0);
toc
x1 = X(:, 1);
x2 = X(:, 2);
subplot(1,2,1), plot(tau, x1), grid
axis square
title('x_1'), xlabel('\tau')
subplot(1,2,2), plot(tau, x2), grid
axis square
title('x_2'), xlabel('\tau')
```

LISTING 6.8   Function File for Example 6.7

```
function dx = ckt6_7(tau, x)
global A B
u = cos(2*pi*tau);
dx = A*x + B*u;
```

Conclusion

In accordance with Equation 6.43, 1 second in time normalized at 10 Hz results in 10 units of τ. Hence, the time vector in *ode23* is specified as [0 10]. It is not necessary to change *t* to *tau* in the main program or in the function file, but it has been done so here for clarity. The results of the simulation are shown in Figure 6.11.

The results of the simulation shown in Figure 6.11 are indeed the same as those in Figure 6.10. The only difference is the normalized horizontal axis in Figure 6.11.

EXERCISE 6.7

Simulate the normalized differential equation from Exercise 6.5. Plot the state variables.

Answer

See Figure 6.12.

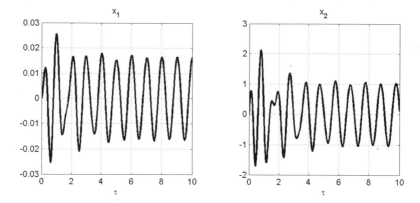

FIGURE 6.11   Simulation results of the normalized system.

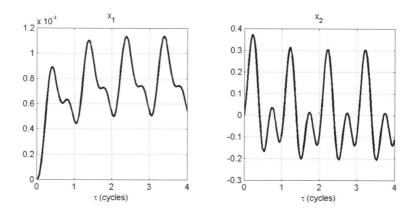

FIGURE 6.12   Solution to Exercise 6.7.

Normalized DC-DC Converter Models

There is little to no difference in the simulation times of the system in Example 6.6 and the normalized system in Example 6.7 because the forcing function changes relatively slowly. The forcing functions in DC-DC converters, however, change quite abruptly and lead to large state derivatives. The normalized DC-DC converter model simulates significantly faster than the time-based model.

The normalized state-space models of the buck, boost, and buck/boost converters are obtained upon substitution of Equation 6.45 into Equations 6.28, 6.35, and 6.42.

Normalized State-Space Model of the Buck Converter

The normalized buck converter model is

$$
\frac{d}{d\tau}\begin{bmatrix} i \\ v \end{bmatrix} = \begin{bmatrix} 0 & -\dfrac{1}{Lf} \\ \dfrac{1}{Cf} & -\dfrac{1}{RCf} \end{bmatrix}\begin{bmatrix} i \\ v \end{bmatrix} + \begin{bmatrix} \dfrac{V_s}{Lf} \\ 0 \end{bmatrix} u. \tag{6.51}
$$

The frequency in the model is the PWM switching frequency. The simulation "time" is the number of integral periods.

Normalized State-Space Model of the Boost Converter

The normalized boost converter model is

$$
\frac{d}{d\tau}\begin{bmatrix} i \\ v \end{bmatrix} = \begin{bmatrix} 0 & -\dfrac{1}{Lf} \\ \dfrac{1}{Cf} & -\dfrac{1}{RCf} \end{bmatrix}\begin{bmatrix} i \\ v \end{bmatrix} + \begin{bmatrix} \dfrac{v}{Lf} & \dfrac{1}{Lf} \\ -\dfrac{i}{Cf} & 0 \end{bmatrix}\begin{bmatrix} u \\ V_s \end{bmatrix}. \tag{6.52}
$$

As the buck converter, division of the model by the PWM switching frequency f results in smaller magnitude derivatives and shorter simulation times.

Normalized State-Space Model of the Buck/Boost Converter

The state-space model of the normalized buck/boost converter is

$$
\frac{d}{d\tau}\begin{bmatrix} i \\ v \end{bmatrix} = \begin{bmatrix} 0 & -\dfrac{1}{Lf} \\ \dfrac{1}{Cf} & -\dfrac{1}{RCf} \end{bmatrix}\begin{bmatrix} i \\ v \end{bmatrix} + \begin{bmatrix} \dfrac{1}{Lf}\left(v+V_s\right) \\ -\dfrac{i}{Cf} \end{bmatrix} u. \tag{6.53}
$$

As the boost converter, the control input matrix is a function of the state variables; however, division by the PWM switching frequency reduces the effect of this dependency.

Simulation of CCM Converters

The simulation process of CCM converts is relatively simple. The binary control variable is switched between zero and one based on whether the current value of τ

falls within the off-time or on-time portion of the switching period. The following examples illustrate the techniques of CCM converter simulation.

EXAMPLE 6.8

Simulate a buck converter with circuit parameters V_s = 12 VDC, V_o = 5 VDC, R = 2.5 Ω, L = 200 μH, C = 47 μF, and f = 5 kHz.

Solution

ON THE CD

The solution programs appear in Listings 6.9 and 6.10 and are available on the CD-ROM as Example6_8.m and buck.m in Chapter 6\Examples and Listings.

LISTING 6.9   Code to Simulate a CCM Buck Converter

```
close all, clear all, clc
global A B D
f = 5000;
R = 2.5;
Vo = 5;
Io = Vo/R;
Vs = 12;
D = Vo/Vs;
L = 200E-6;
C = 47E-6;
A = [0 -1/(L*f); 1/(C*f) -1/(f*R*C)];
B = [Vs/(f*L); 0];
x0 = [Io; Vo];
tf = 10;
tic
[t, X] = ode23('buck', [0 tf], x0);
toc
IL = X(:, 1);
VC = X(:, 2);
subplot(2,1,1), plot(t, IL), grid
title('Inductor Current')
subplot(2,1,2), plot(t, VC), grid
axis([0 tf 0 10])
title('Output Voltage')
xlabel('cycles')
```

LISTING 6.10   Function File for Simulation of the Buck Converter

```
function dx = buck(t, x)
global A B D
u = 0.5*(1 - sign(t - fix(t) - D));
dx = A*x + B*u;
```

Conclusion

The initial conditions of the simulation are chosen as the load current and the desired output voltage in order for the system to reach the steady-state condition in the shortest amount of time. Although the variable name t is used in the files, the independent variable is the normalized τ, which represents the number of cycles of the PWM switching frequency. The specified simulation interval of [0 10] thus presents a time interval of 10 cycles, or 10/5000 = 2 ms.

The binary switching control u is based on MATLAB's *sign* function, which is a numerical implementation of the *signum* function, $\sigma(x)$, illustrated in Figure 6.13.

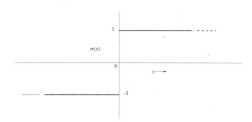

FIGURE 6.13   The signum function.

As shown in Figure 6.13, $\sigma(x) = 1$ when $x > 1$ and $\sigma(x) = -1$ when $x < 0$. When $x = 0$, $\sigma(x)$ is undefined. The *sign* function works similarly, except that when the argument is zero, *sign* returns a value of zero.

As used in the function file of Listing 6.10, *sign* returns a value of 1, −1, or 0 based on whether the quantity $t - fix(t) - D$ is positive, negative, or zero, respectively. Variable t represents the current value of τ. The `fix` function truncates any fractional part of t and leaves only the integer part. The quantity $t - fix(t)$ is thus the fractional part of τ. For example, if $t = 5.734$, then $fix(t) = 5$, and $t - fix(t) = 0.734$. Since t represents cycles, the quantity $t - fix(t)$ represents the fraction of a cycle. If this fraction is greater than the duty ratio, then the quantity $t - fix(t) - D$ is positive, and $u = 0.5*(1 - 1) = 0$. This value of t corresponds to a point in the off-time interval of the switching period. The primary switch is therefore open and the inductor is disconnected from the source. If the cycle fraction is less than the duty ratio, then the quantity $t - fix(t) - D$ is negative, and the binary switch becomes $u = 0.5^*[1 - (-1)] = 1$. This value of t falls within the on-time interval of the switching period during which the inductor is connected to the source. The results of the simulation, the inductor current, and capacitor voltage (also the load voltage) are plotted in Figure 6.14.

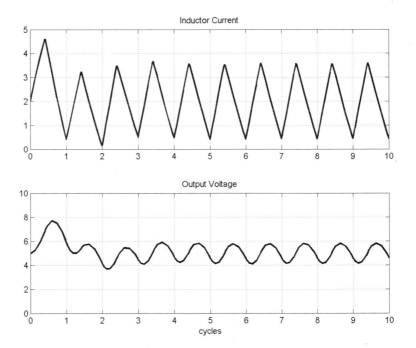

FIGURE 6.14   CCM buck converter simulation.

The CCM inductance of the converter of this example is 146 μH. The actual inductance used, 200 μH, represents a λ-ratio of approximately 1.37. The minimum and maximum inductor currents calculated in terms of the λ-ratio are

$$I_{min} = \frac{\lambda - 1}{\lambda} I_o = \frac{1.37 - 1}{1.37}(2.0) = 0.54 \text{ A}$$

and

$$I_{max} = \frac{\lambda + 1}{\lambda} I_o = \frac{1.37 + 1}{1.37}(2.0) = 3.46 \text{ A}.$$

These values agree remarkably well with the steady-state inductor current shown in Figure 6.14.

The steady-state peak-to-peak ripple voltage from Figure 6.13 is approximately 1.7 V and corresponds to a ripple content of 1.7/5.0 = 0.34, or 34%. The ripple computed from the circuit parameters is

$$r = \frac{1-D}{8LCf^2} = \frac{1-5/12}{(8)(200E-6)(47E-6)(5E3)^2} = 0.31 \text{ or } 31\%.$$

In practice, much larger capacitances are used to significantly reduce the ripple.

The simulation results are in excellent agreement with the buck converter theory developed in Chapter 4.

EXERCISE 6.8

Change the value of the capacitance in the program of Listing 6.9 to 470 µf and determine the peak-to-peak ripple voltage with MATLAB's *max* and *min* functions applied to the steady-state output voltage. Simulate 40 cycles of the PWM switching frequency.

Answer

$\Delta V_o \approx 0.2$ V.

EXAMPLE 6.9

Simulate a boost converter with circuit parameters $V_s = 12$ VDC, $V_o = 28$ VDC, R = 8.0 Ω, $L = 150$ µH, C = 100 µF, and $f = 5$ kHz.

Solution

ON THE CD

The solution programs are provided in Listings 6.11 and 6.12 and are available on the CD-ROM as Example6_9.m and boost.m in Chapter 6\Examples and Listings.

LISTING 6.11 Code to Simulate a CCM Boost Converter

```
close all, clear all, clc
global A D Cf Lf Vs
f = 5000;
R = 8;
Vo = 28;
Io = Vo/R;
Vs = 12;
D = 1 - Vs/Vo;
```

```
L = 150E-6;
C = 100E-6;
Lf = L*f;
Cf = C*f;
A = [0 -1/Lf; 1/Cf -1/(R*Cf)];
x0 = [D*Io; Vo];
tf = 20;
tic
[t, X] = ode23('boost', [0 tf], x0);
toc
IL = X(:, 1);
VC = X(:, 2);
subplot(2,1,1), plot(t, IL), grid
axis([0 tf 0 15])
title('Inductor Current')
subplot(2,1,2), plot(t, VC), grid
axis([0 tf 0 35])
title('Output Voltage')
xlabel('cycles')
```

LISTING 6.12   Function File for Simulation of the Boost Converter

```
function dx = boost(t, x)
global A D Cf Lf Vs
iL = x(1);
vC = x(2);
B = [vC/Lf 1/Lf; -iL/Cf 0];
u = [0.5*(1 - sign(t - fix(t) - D)); Vs];
dx = A*x + B*u;
```

Conclusion

The results of the simulation are shown in Figure 6.15.

The minimum inductance required for CCM operation of the boost converter in this example is

$$L_{\text{CCM}} = \frac{D(1-D)^2 R}{2f} = \frac{(0.57)(0.43)^2(8)}{(2)(5000)} = 84 \ \mu\text{H}.$$

The control input matrix does not appear in the *global* statement of Listings 6.11 and 6.12 because it is not constant; it must be computed in the same function file as the state derivative. In the function file, the inductor current and capacitor voltage are extracted from the state vector, as a matter of convenience, before computation of the control input matrix.

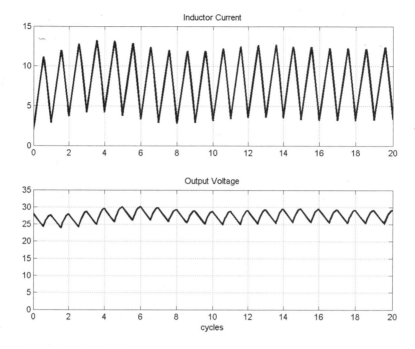

FIGURE 6.15   CCM boost converter simulation.

The initial condition for the inductor current is set to DI_o instead of I_o because the load-connected duty ratio of the boost converter is less than unity.

EXERCISE 6.9

Simulate the boost converter of Example 6.9 with a load resistance of 4 Ω. Determine the peak steady-state inductor current.

Answer

$I_{max} = 20$ A.

EXAMPLE 6.10

Simulate a buck/boost converter with circuit parameters $V_s = 12$ VDC, $V_o = -12$ VDC, R = 4.0 Ω, L = 300 μH, C = 75 μF, and $f = 10$ kHz.

Solution

ON THE CD

The solution code is provided in Listings 6.13 and 6.14 and also on the CD-ROM as Example6_10.m and bkboost.m in Chapter 6/Examples.

LISTING 6.13   Code to Simulate a CCM Buck/Boost Converter

```
close all, clear all, clc
global A D Cf Lf Vs
f = 10000;
R = 4;
Vo = 12;
Io = Vo/R;
Vs = 12;
D = Vo/(Vo + Vs);
L = 300E-6;
C = 75E-6;
Lf = L*f;
Cf = C*f;
A = [0 -1/Lf; 1/Cf -1/(R*Cf)];
x0 = [D*Io; Vo];
tf = 20;
tic
[t, X] = ode23('bkboost', [0 tf], x0);
toc
IL = X(:, 1);
VC = X(:, 2);
subplot(2,1,1), plot(t, IL), grid
axis([0 tf 0 5])
title('Inductor Current')
subplot(2,1,2), plot(t, VC), grid
axis([0 tf 0 20])
title('Output Voltage')
xlabel('cycles')
```

LISTING 6.14   Function File for Simulation of the Buck/Boost Converter

```
function dx = bkboost(t, x)
global A D Cf Lf Vs
iL = x(1);
vC = x(2);
B = [(vC + Vs)/Lf; -iL/Cf];
u = 0.5*(1 - sign(t - fix(t) - D));
dx = A*x + B*u;
```

Conclusion

The simulation results are shown in Figure 6.16.

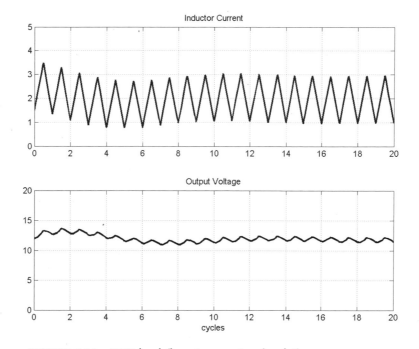

FIGURE 6.16   CCM buck/boost converter simulation.

The minimum required CCM inductance for the converter in this example is

$$L_{\text{CCM}} = \frac{\left(1-D\right)^2 R}{2f} = \frac{(0.5)^2(4)}{(2)(10\text{E}3)} = 50 \ \mu\text{H}.$$

As with the boost converter, the control input matrix must be computed in the function file. Also, the initial inductor current is DI_o because of the load-connected duty ratio.

The output voltage appears positive in Figure 6.16 because of the polarity indicated in the circuit schematic when the design equations were developed for the buck/boost converter. The output voltage is actually of opposite polarity than that of the source voltage.

EXERCISE 6.10

Simulate the buck/boost converter in Example 6.10 with a load resistance of 2 Ω and determine the peak inductor current.

Answer

$I_{\max} = 5$ A.

Conclusion

The simulation results reveal the dynamic behavior of CCM converters and confirm the theory developed in previous chapters. The inductor current waveform has the level-shifted triangular waveform typical of CCM converters.

The simulation results of the buck converter reveal a common problem with some DC-DC converters: significant voltage over-shoot at start-up. This overshoot could damage equipment supplied by the converter. In such over-voltage transients, voltage suppression techniques are employed to absorb the potentially harmful energy.

Simulation of DCM Converters

The simulation process of DCM converters is not quite as direct as that of CCM converters. In DCM, the inductor current must be extinguished before the start of the subsequent switching period. Consequently, to simulate the effect of the diode in the circuit, any computed negative values of current must be replaced by zero in the off-time interval. The simulation thus requires a cycle-by-cycle computation followed by a postprocessing of data to include the diode behavior. Additionally, the initial conditions of each cycle simulation must be reset so that the final capacitor voltage and zero inductor current of the present cycle become the initial capacitor voltage and initial inductor current of the next cycle.

The use of *ode23* also presents a problem to DCM simulation. The nonconstant step length leads to inconsistent inductor current extinction times. A constant step-length Runge-Kutta algorithm is introduced as an alternative to *ode23*.

Runge-Kutta Algorithm with Constant Step Length

As stated previously, *ode23* adjusts the step length of the algorithm to improve numerical accuracy. An alternative to an adjustable step length is to use a smaller constant step length to maintain accuracy and preclude the problems caused by a variable step length. The disadvantage to this alternative is longer simulation times. A constant step-length Runge-Kutta algorithm is implemented in the MATLAB function *rk4* presented in Listing 6.15. Function rk4.m is available on the CD-ROM in the Toolbox subfolder of Chapter 6.

ON THE CD

LISTING 6.15   Constant Step-Length Runge-Kutta Algorithm

```
function [t, X] = rk4(mfile, tvec, X0, N)
t0 = tvec(1);
tf = tvec(2);
h = (tf - t0)/N;
t = linspace(t0,tf,N)';
X(:,1) = X0;
for q = 1:N - 1,
    tq = t(q);
    Xq = X(:,q);
    F1 = h*feval(mfile, tq, Xq);
    tq2 = tq + h/2;
    F2 = h*feval(mfile, tq2, Xq + F1/2);
    F3 = h*feval(mfile, tq2, Xq + F2/2);
    F4 = h*feval(mfile, tq + h, Xq + F3);
    X(:, q + 1) = Xq + ([1 2 2 1]*[F1 F2 F3 F4]'/6)';
end
X = X';
```

The function arguments are in the same order and structure as *ode23*, with the exception that the number of computed points is specified by the last argument *N*. The structure of the output of *rk4* is identical to *ode23*. The next example illustrates the use of *rk4* in a simulation.

Example 6.11

Simulate the normalized system of Exercise 6.5 with the *rk4* function.

Solution

ON THE CD

The solution programs are presented in Listings 6.16 and 6.17 and are available on the CD-ROM in folder Chapter 6\Examples and Listings as files Example6_11.m and ckt6_11.m.

LISTING 6.16   Simulation with a Constant Step-Length Runge-Kutta Algorithm

```
close all, clear all, clc
global A B
A = [0 1/1000; -1 -2];
B = [0; 1];
X0 = [0; 0];
```

```
tic
[t, X] = rk4('ckt6_11', [0 4], X0, 500);
toc
x1 = X(:, 1);
x2 = X(:, 2);
subplot(1,2,1), plot(t, x1), grid
axis square
title('x_1'), xlabel('\tau (cycles)')
subplot(1,2,2), plot(t, x2), grid
axis square
title('x_2'), xlabel('\tau (cycles)')
```

LISTING 6.17   Function File for Example 6.11

```
function dx = ckt6_11(t, x)
global A B
u = cos(2*pi*t) + 2*sin(4*pi*t);
dx = A*x + B*u;
```

Conclusion

Listings 6.16 and 6.17 show that the only difference between the use of *rk4* and *ode23* is the specification of the number of computed points in *rk4*.

EXERCISE 6.11

Use *rk4* to simulate the CCM buck converter of Example 6.8.

The following examples illustrate the technique of DCM converter simulation.

EXAMPLE 6.12

Simulate the buck converter in Example 6.8 with $L = 50\ \mu H$ and $C = 470\ \mu F$.

Solution

ON THE CD

The solution program is provided in Listing 6.18 and as Example6_12.m in Chapter 6\Examples and Listings on the CD-ROM.

LISTING 6.18   Code to Simulate a DCM Buck Converter

```
close all, clear all, clc
global A B D
N = 100; % Number of computed points per cycle
h = 1/N; % step-length
f = 7000;
```

```
R = 2.5;
Vo = 5;
Io = Vo/R;
Vs = 12;
L = 50E-6;
Lmin = lccm(Vs, Vo, Io, f, 'buck');
lmda = L/Lmin;
D = Vo/Vs*sqrt(lmda);
C = 470E-6;
A = [0 -1/(L*f); 1/(C*f) -1/(f*R*C)];
B = [Vs/(f*L); 0];
x0 = [0; Vo];
VC = [];
IL = [];
tau = [];
ti = 0;
tf = 1 - h;
tic
for T = 1:10,
    [t, X] = rk4('buck', [ti tf], x0, N);
    iL = X(:, 1);
    vC = X(:, 2);
    m = length(iL);
    n = fix(D*m);
    for q = n:m,
        if iL(q) < 0,
            iL(q) = 0; % zero the negative values of inductor current
        end
    end
    IL = [IL; iL];
    VC = [VC; vC];
    tau = [tau; t];
    x0 = [0; vC(m)]; % Reset the initial conditions
    ti = t(m) + h;
    tf = T + 1 - h;
end
toc
subplot(2,1,1), plot(tau, IL, 'k'), grid
axis([0 T -5 10])
title('Inductor Current')
subplot(2,1,2), plot(tau, VC, 'k'), grid
axis([0 T 0 10])
title('Output Voltage')
xlabel('cycles')
```

Conclusion

The *buck* function specified in *rk4* is the same code as in Listing 6.10. The cycle-by-cycle simulation requires *rk4* to be repeated for as many cycles as specified by variable *T* in the first *for* statement. Once the data for one cycle is computed, the state variables are extracted into temporary variables *iL* and *vC*. The second *for* loop searches for negative values of inductor current within the off-time interval and replaces those values with zero. The processed inductor current and computed capacitor voltage are then stored in vectors **IL** and **VC**, respectively, while the normalized time vector **t** is stored in vector **tau**. The initial conditions are subsequently reset to zero inductor current and the final value of capacitor voltage from the present cycle. The initial time is set to the beginning time of the next cycle, and the cycle counter is advanced to the next period. The simulation results are shown in Figure 6.17.

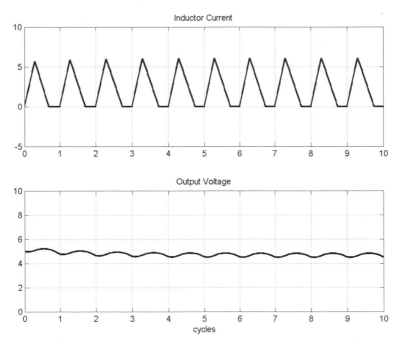

FIGURE 6.17 DCM buck converter simulation.

The simulation results show that the converter outputs 5 VDC with the duty ratio set to the generalized DCM duty ratio developed in Chapter 5.

The program in Listing 6.18 uses the *lccm* function to compute the CCM inductance. Function *lccm.m* is located in the Toolbox subfolder of Chapter 6 on the

ON THE CD CD-ROM. The function computes the CCM inductance for the buck, boost, and buck/boost converters. The default computation is for the buck/boost converter; otherwise the strings *'buck'* or *'boost'* specify the respective converter.

EXERCISE 6.12

Simulate the DCM buck converter from Example 6.12 with a switching frequency of 5 kHz and observe the output voltage and the ripple voltage.

EXAMPLE 6.13

Simulate the boost converter in Example 6.9 with $L = 20\ \mu H$, $C = 100\ \mu F$, and $f = 15$ kHz.

Solution

ON THE CD The solution program is provided in Listing 6.19 and as Example6_13.m in Chapter 6\Examples and Listings on the CD-ROM.

LISTING 6.19   Program to Simulate the DCM Boost Converter

```
close all, clear all, clc
global A D Cf Lf Vs
N = 100;
h = 1/N;
f = 15000;
R = 8;
Vo = 28;
Io = Vo/R;
Vs = 12;
L = 20E-6;
Lmin = lccm(Vs, Vo, Io, f, 'boost');
lmda = L/Lmin;
D = (1 - Vs/Vo)*sqrt(lmda);
C = 100E-6;
Lf = L*f;
Cf = C*f;
A = [0 -1/Lf; 1/Cf -1/(R*Cf)];
x0 = [0; Vo];
VC = [];
IL = [];
tau = [];
ti = 0;
```

```
tf = 1 - h;
tic
for T = 1:10,
    [t, X] = rk4('boost', [ti tf], x0, N);
    iL = X(:, 1);
    vC = X(:, 2);
    m = length(iL);
    n = fix(D*m);
    for q = n:m,
        if iL(q) < 0,
            iL(q) = 0;
        end
    end
    IL = [IL; iL];
    VC = [VC; vC];
    tau = [tau; t];
    x0 = [0; vC(m)];
    ti = t(m) + h;
    tf = T + 1 - h;
end
toc
subplot(2,1,1), plot(tau, IL, 'k'), grid
axis([0 T -5 30])
title('Inductor Current')
subplot(2,1,2), plot(tau, VC, 'k'), grid
axis([0 T 0 40])
title('Output Voltage')
xlabel('cycles')
```

Conclusion

The *boost* function specified in *rk4* is the same code as in Listing 6.12. The results of the simulation are shown in Figure 6.18.

EXERCISE 6.13

Simulate the boost converter at a switching frequency of 10 kHz and observe the output voltage and the ripple voltage.

EXAMPLE 6.14

Simulate the buck/boost converter in Example 6.10 with L = 10 μH, C = 220 μF, and f = 20 kHz.

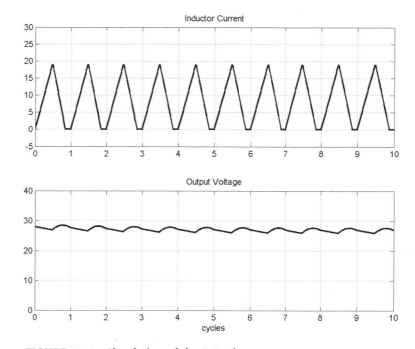

FIGURE 6.18 Simulation of the DCM boost converter.

Solution

The solution program is provided in Listing 6.20 and as Example6_14.m on the CD-ROM in the Examples and Listings subfolder of Chapter 6.

LISTING 6.20 Program to Simulate the DCM Buck/Boost Converter

```
close all, clear all, clc
global A D Cf Lf Vs
N = 100;
h = 1/N;
f = 20000;
R = 4;
Vo = 12;
Io = Vo/R;
Vs = 12;
L = 10E-6;
Lmin = lccm(Vs, Vo, Io, f, '');
lmda = L/Lmin;
```

```
D = Vo/(Vo + Vs)*sqrt(lmda);
C = 220E-6;
Lf = L*f;
Cf = C*f;
A = [0 -1/Lf; 1/Cf -1/(R*Cf)];
x0 = [0; Vo];
VC = [];
IL = [];
tau = [];
ti = 0;
tf = 1 - h;
tic
for T = 1:10,
    [t, X] = rk4('bkboost', [ti tf], x0, N);
    iL = X(:, 1);
    vC = X(:, 2);
    m = length(iL);
    n = fix(D*m);
    for q = n:m,
        if iL(q) < 0,
            iL(q) = 0;
        end
    end
    IL = [IL; iL];
    VC = [VC; vC];
    tau = [tau; t];
    x0 = [0; vC(m)];
    ti = t(m) + h;
    tf = T + 1 - h;
end
toc
subplot(2,1,1), plot(tau, IL, 'k'), grid
axis([0 T -5 30])
title('Inductor Current')
subplot(2,1,2), plot(tau, VC, 'k'), grid
axis([0 T 0 20])
title('Output Voltage')
xlabel('cycles')
```

Conclusion

The *bkboost* function specified in *rk4* is the same code as in Listing 6.14. The simulation results are shown in Figure 6.19.

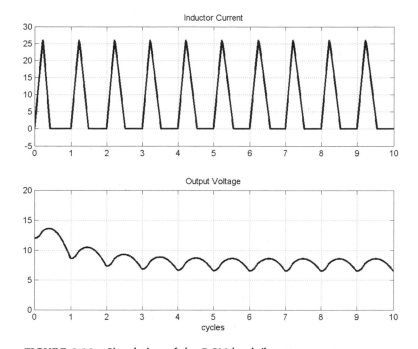

FIGURE 6.19 Simulation of the DCM buck/boost converter.

EXERCISE 6.14

Simulate the buck/boost converter from Example 6.14 with an inductance of 5 μH and observe the output voltage and output ripple voltage.

CHAPTER SUMMARY

MATLAB proves to be an excellent platform for the simulation of DC-DC converters. Unlike PSpice®, no circuit schematic or cryptic netlist is required for the simulation—only the circuit equations derived by application of KCL and KVL and formed into a normalized state-space model.

The simulation methods for both CCM and DCM operations presented are applicable to virtually any DC-DC converter.

THE MATLAB TOOLBOX

The functions listed below are located in subfolder Toolbox of Chapter 6 on the CD-ROM

BUCK

Function *buck.m* computes the state derivative of the normalized state-space model of the buck converter. See Examples 6.8 and 6.12 for syntax and use with *ode23* and *rk4*.

BOOST

Function *boost.m* computes the state derivative of the normalized state-space model of the boost converter. See Examples 6.9 and 6.13 for syntax and use with *ode23* and *rk4*.

BKBOOST

Function *bkboost.m* computes the state derivative of the normalized state-space model of the buck/boost converter. See Examples 6.10 and 6.14 for syntax and use with *ode23* and *rk4*.

LCCM

Function *lccm.m* computes the CCM inductance for the buck, boost, and buck/boost converters from the input and output voltages, the load current, and the PWM switching frequency.

Syntax

```
Lmin = lccm(Vs, Vo, Io, f, 'converter')
```

String variable '*converter*' is 'buck' or 'boost' for the respective converter; otherwise the CCM inductance for the buck/boost converter is computed.

Example

```
Lmin = lccm(12, 5, 2, 15E3, 'buck')
```

RK4

Function *rk4.m* computes the integral of the state derivative of a state-space model by use of a fourth-order, constant step-length, Runge-Kutta algorithm.

Syntax

```
[t, X] = rk4('file', [t0 tf], X0, N)
```

String *'file'* is the filename with *.m* extension that contains the state derivative. Vector [**t0 tf**] is the time interval over which the state derivative is integrated, vector **X0** is the initial state vector, and N is the number of computed points within the time interval. The step length is

$$\frac{t_f - t_0}{N}.$$

PROBLEMS

Problem 1

Simulate the buck converter of Problem 3, Chapter 4.

Problem 2

Simulate the buck converter of Problem 4, Chapter 4.

Problem 3

Simulate the boost converter of Problem 3, Chapter 5.

Problem 4

Simulate the boost converter of Problem 4, Chapter 5.

Problem 5

Simulate the buck/boost converter of Problem 7, Chapter 5.

Problem 6

Simulate the buck/boost converter of Problem 8, Chapter 5.

Problem 7

Develop the normalized state-space model of the Watkins-Johnson converter from Chapter 5.

Problem 8

Develop the normalized state-space model of the inverse Watkins-Johnson converter from Chapter 5.

Problem 9

Develop the normalized state-space model of the bridge converter from Chapter 5.

Problem 10

Develop the normalized state-space model of the current-fed converter from Chapter 5.

Problem 11

Simulate the Watkins-Johnson converter with the same circuit parameters as the boost converter in Problem 1, Chapter 5.

Problem 12

Simulate the inverse Watkins-Johnson converter with the same circuit parameters as the boost converter in Problem 3, Chapter 5.

7 Inverters: Converting DC to AC

In This Chapter

INTRODUCTION

An inverter is used to generate an AC voltage from a DC source. Typically, a transformer is employed to step the voltage up significantly above the magnitude of the DC voltage. The DC source polarity is alternately switched across the transformer primary, which produces an alternating voltage at the transformer secondary, where the load is connected.

If the AC load is purely resistive, the actual waveform shape produced at the secondary is not critical. Examples of purely resistive AC loads are incandescent lights and heating elements. If, for example, a 120-VAC coffee pot is to be powered from an automobile starting battery, the inverter must generate the necessary RMS voltage that results in the average power required by the appliance. Since the desired effect is heat, the waveform shape across the load is not important; it need not even be AC.

For some loads, however, the waveform shape is essential. An AC induction motor, for example, is designed specifically to operate from a sinusoidal source. Motor torque variations and additional heat losses result if the phase current waveforms deviate from the sinusoidal shape.

The focus of this chapter is inverters that employ transformers to generate high-voltage AC from low-voltage DC. The presentation begins with a review of the fundamentals of the AC transformer and concludes with inverter simulation techniques in MATLAB.

AC TRANSFORMERS

The operation of the AC transformer is based on Michael Faraday's law of induction: a changing current in a coil induces a current in a second coil that is in close proximity to the first. The key to induction is a changing current: the current must change with time in order to induce current elsewhere. A constant current, such as that produced by a DC source, cannot be transformed.

The Ideal Transformer

A schematic of an *ideal*, two-winding transformer is illustrated in Figure 7.1.

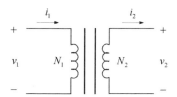

FIGURE 7.1   Ideal, two-winding transformer.

The heavy lines shown between the coils in Figure 7.1 represent the core of the transformer, which is typically iron. The core of the ideal transformer is infinitely permeable. The windings of an ideal transformer are perfectly coupled and suffer no losses. If the primary winding is composed of N_1 turns of wire and the secondary with N_2 turns of wire, the ratio of the primary voltage to the secondary voltage is equal to the ratio of the primary turns to the secondary turns:

$$\frac{v_1}{v_2} = \frac{N_1}{N_2}. \tag{7.1}$$

The ratio N_1/N_2 is called the *turns-ratio* of the transformer and is notated as $N_1{:}N_2$ or as

$$a = \frac{N_1}{N_2}. \tag{7.2}$$

The secondary voltage in terms of the primary voltage and the turns-ratio is

$$v_2 = \frac{v_1}{a}. \tag{7.3}$$

If $a > 1$, the transformer is called a buck, or step-down transformer. If $a < 1$, the transformer is a boost, or step-up transformer. The ratio of the primary and secondary currents is related inversely to the transformer turns ratio:

$$\frac{i_1}{i_2} = \frac{1}{a}. \tag{7.4}$$

EXAMPLE 7.1

A transformer with a turns ratio of $a = 0.25$ draws 10 A from a 120 VAC source. Determine the secondary voltage and current.

Solution
Directly from Equations 7.3 and 7.4, the secondary voltage is

$$v_2 = \frac{v_1}{a} = \frac{120}{0.25} = 480 \text{ VAC.}$$

and the secondary current is

$$i_2 = ai_1 = (0.25)(10) = 2.5 \text{ A.}$$

Conclusion
Since $a < 1$, the transformer is a boost-type. The output voltage is four times the input voltage, and the secondary current is one-fourth of the primary current.

EXERCISE 7.1

Repeat Example 7.1 for a turns ratio of $a = 2$.

Answers

$v_2 = 60$ VAC, $i_2 = 20$ A.

If the ideal transformer has a secondary-connected impedance Z_2, the load impedance in terms of the secondary voltage and secondary current is expressed as

$$Z_2 = \frac{v_2}{i_2}. \tag{7.5}$$

The equivalent impedance at the primary terminals is

$$Z_1 = \frac{v_1}{i_1}. \tag{7.6}$$

Substitution of the relationships of Equations 7.3 and 7.4 into Equation 7.6 results in

$$Z_1 = a^2 \frac{v_2}{i_2}. \tag{7.7}$$

In terms of the secondary impedance expressed by Equation 7.5, the impedance at the primary terminals is

$$Z_1 = a^2 Z_2. \tag{7.8}$$

Equation 7.8 reveals that the equivalent impedance at the primary winding is equal to the square of the turns ratio multiplied by the load impedance connected to the secondary winding. Thus, an ideal transformer with a load impedance may be replaced by an equivalent impedance *referred* to the primary winding. This impedance referral is illustrated in Figure 7.2.

The referred impedance of a step-down transformer is greater than the original load impedance, while the referred impedance of a step-up transformer is less. The utility of impedance referral is illustrated in the next example.

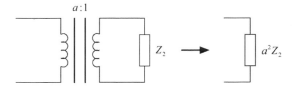

FIGURE 7.2   Impedance referral from secondary to primary.

EXAMPLE 7.2

Determine the power delivered by the source to the load impedance of the circuit shown in Figure 7.3 with $V_s = 240$ VAC.

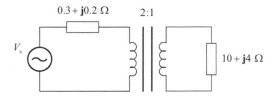

FIGURE 7.3   Circuit from Example 7.2.

Solution

ON THE CD

The solution is provided by the code in Listing 7.1. The program is also available as Example7_2.m in Chapter 7\Examples and Listings on the CD-ROM.

LISTING 7.1   Solution Code for Example 7.2

```
close all, clear all, clc
Vs = 240;
a = 2;
ZL = 10 + j*4;
Zs = 0.3 + j*0.2;
Zr = a^2 * ZL;
Ze = Zs + Zr;
Is = Vs/abs(Ze);
P = Is^2 * real(Zr)
```

Conclusion

Once the impedance is referred to the primary, it is summed with the primary series impedance to form the equivalent impedance seen by the source. The RMS current drawn from the source is the RMS source voltage divided by the magnitude of the equivalent impedance. The average power delivered to the load, $P = 1.2$ kW, is dissipated in the real part of the referred impedance.

EXERCISE 7.2

Determine the power delivered to the load of the circuit shown in Figure 7.4 with $V_s = 480$ VAC.

Answer

8.2 kW.

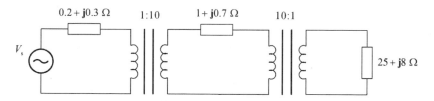

FIGURE 7.4   Circuit for Exercise 7.2.

Physical Transformer Model

A physical transformer incurs losses both in the windings from series resistance and in the iron because of currents induced in the core. These losses are modeled by the addition of series and parallel resistive and reactive elements to the ideal transformer model. The schematic of the physical transformer is illustrated in Figure 7.5.

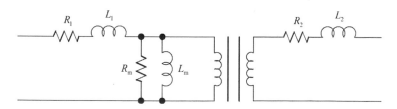

FIGURE 7.5   Physical transformer model.

Series elements R_1, L_1 and R_2, L_2 represent, respectively, the resistance and inductance of the primary and secondary windings. The series elements model the voltage drops that occur in the physical transformer; the resistances model the power losses in the windings. Parallel element R_m models the losses that occur in the core. Parallel element L_m is called the *magnetizing inductance* and represents the current required to permeate the core with a magnetic field. In the ideal transformer model the series elements are zero, while the parallel elements are infinite.

The physical transformer model is crucial to the design of inverters that must output a sinusoidal waveform. Indeed, the loss elements actually assist in the waveshaping process, as they impede the higher-frequency components. The physical model also determines the overall efficiency of the circuit. The next example illustrates how the physical transformer model affects the efficiency computation of a circuit.

EXAMPLE 7.3

Compute the efficiency of the circuit shown in Figure 7.6 with (a) the ideal transformer model and (b) the physical transformer model with $r_1 = 0.1\ \Omega$, $L_1 = 50\ \mu H$, $r_m = 1\ k\Omega$, $L_1 = 0.8\ H$, $r_2 = 0.02\ \Omega$, and $L_2 = 10\ \mu H$. The source voltage is $V_s = 120$ VAC at 60 Hz, the turns ratio is $a = 5$, and the load impedance is $Z_L = 1 + j0.3\ \Omega$.

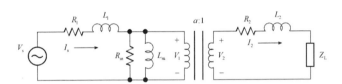

FIGURE 7.6   Circuit for Example 7.3.

Solution

With the ideal transformer model, there are no computations to perform; the efficiency is 100%. Although computation of average load power is simple with the ideal transformer, the ideal model is not suitable for efficiency computations.

With the physical model, the load impedance and secondary winding elements are referred to the primary winding to facilitate the efficiency calculation. The solution is provided by the code in Listing 7.2 and is available as Example7_3.m in Chapter 7\Examples and Listings on the CD-ROM.

ON THE CD

LISTING 7.2   Code to Compute the Efficiency of the Circuit in Figure 7.6

```
close all, clear all, clc
Vs = 120;
a = 5;
uH = 1E-6;
w = 120*pi;
R1 = 0.1;
R2 = 0.02;
Rm = 1000;
L1 = 50*uH;
L2 = 10*uH;
Lm = 0.8;
X1 = w*L1;
X2 = w*L2;
Xm = w*Lm;
ZL = 1 + j*0.3;
Z2 = R2 + j*X2 + ZL;
Zr = a^2 * Z2;
Zp = 1/(1/Rm + 1/(j*Xm) + 1/Zr);
Zs = R1 + j*X1;
V1 = abs(Vs*Zp/(Zp + Zs));
V2 = V1/a;
I2 = abs(V2/Z2);
PL = I2^2 * real(ZL);
Z1 = Zs + Zp;
Is = abs(Vs/Z1);
P1 = Is^2 * R1;
Pm = V1^2 / Rm;
P2 = I2^2 * R2;
eta = PL/(PL + P1 + Pm + P2)
```

Conclusion

Once the secondary winding impedance and the load impedance are referred to the primary, the referred impedance is combined in parallel with the magnetizing impedance. The primary voltage of the ideal transformer is found by voltage division. The secondary voltage of the ideal transformer is determined from the turns ratio; the secondary current is determined from the secondary voltage and the total secondary impedance. The power losses are determined from the loss elements and their respective RMS values of voltage or current. The computed value of efficiency is $\eta = 95\%$.

EXERCISE 7.3

The question often arises as to whether it is possible to use a step-down transformer as a step-up transformer simply by "turning the transformer around," that is, by using the secondary winding as the primary and vice versa. It is possible as long as the voltage and current ratings of the windings are not exceeded. The output voltage, however, tends to be lower than expected because of the larger impedance of the primary winding.

Compute the output voltage of the circuit shown in Figure 7.7, in which the same transformer from Example 7.3 is used with a voltage source of 24 VAC applied to the secondary winding and a load impedance of $Z_L = 25 + j7.5 \ \Omega$ is connected to the primary winding.

Answer

$V_L = 117$ VAC.

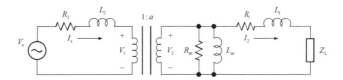

FIGURE 7.7   Circuit for Exercise 7.3.

THE INVERTER CIRCUIT

The functional schematic of the *full-bridge,* or *H-bridge,* inverter is illustrated in Figure 7.8.

The alternating voltage is produced when switches SW1 and SW3 are closed for half the switching period while SW2 and SW4 are open. In the second half-period SW1 and SW3 are open while SW2 and SW4 are closed. The load voltage is thus

$$V_L = \begin{cases} V_b, & 0 < \theta < \pi \\ -V_b, & \pi < \theta < 2\pi \end{cases}. \tag{7.9}$$

The load voltage is illustrated in Figure 7.9.

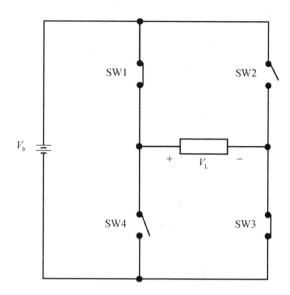

FIGURE 7.8   Full-bridge inverter circuit.

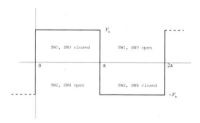

FIGURE 7.9   Load voltage of the H-bridge inverter.

Square-Wave Inverter

The circuit in Figure 7.8 is often referred to as a *square-wave* inverter because of the rectangular wave shape of the load voltage produced by the switching sequence. If the load of the inverter is purely resistive, the load current is simply Equation 7.9 divided by the load resistance

$$i_L = \begin{cases} \dfrac{V_b}{R}, & 0 < \theta < \pi \\[2ex] -\dfrac{V_b}{R}, & \pi < \theta < 2\pi \end{cases}, \qquad (7.10)$$

and the power absorbed in the load is

$$P = \frac{V_b^2}{R}.$$ (7.11)

Square-Wave Inverter with Inductive Load

In practice, the load impedance of the square-wave inverter is significantly inductive, especially when the load is the primary winding of a step-up transformer. Inductive loads require diodes in the circuit to provide a conductive path for the energy stored in the inductance when the switches are opened. The diodes protect the switches from large voltage transients that would otherwise result from a sudden interruption of the load current. Figure 7.10 shows the full-bridge inverter with protective diodes placed across the switches.

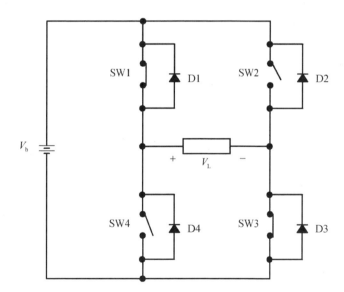

FIGURE 7.10   Inverter circuit with protective diodes.

If the inductive load is modeled as a series resistance and inductance, application of KVL to the circuit yields the differential equation

$$V_b = \omega L \frac{di}{d\theta} + iR.$$ (7.12)

In terms of the quality factor of the load, Equation 7.12 becomes

$$\frac{di}{d\theta} + \frac{i}{q} = \frac{V_b}{qR}. \tag{7.13}$$

With switch pair SW1-SW3 closed and SW2-SW4 open, the solution of Equation 7.13 is

$$i_1(\theta) = A_1 e^{-\theta/q} + \frac{V_b}{R}, \; 0 \le \theta \le \pi, \tag{7.14}$$

in which A_1 is a constant determined by the boundary conditions. When $\theta = \pi$, switch pair SW1-SW3 opens and SW2-SW4 closes, and the load current becomes

$$i_2(\theta) = A_2 e^{-(\theta-\pi)/q} - \frac{V_b}{R}, \; \pi \le \theta \le 2\pi. \tag{7.15}$$

Since the inductor current cannot change instantaneously, evaluation of Equation 7.14 at the beginning of the period is equal to Equation 7.15 evaluated at the end of the period. For the same reason, Equations 7.14 and 7.15 are also equal at the midpoint of the period. The boundary conditions, therefore, are

$$i_1(0) = i_2(2\pi), \tag{7.16}$$

and

$$i_1(\pi) = i_2(\pi). \tag{7.17}$$

Application of the first boundary condition to Equations 7.14 and 7.15 results in

$$A_1 + \frac{V_b}{R} = A_2 e^{-\pi/q} - \frac{V_b}{R}. \tag{7.18}$$

Application of the second boundary condition produces

$$A_1 e^{-\pi/q} + \frac{V_b}{R} = A_2 - \frac{V_b}{R}. \tag{7.19}$$

The solutions of Equations 7.18 and 7.19 for A_1 and A_2 are

$$A_1 = -\frac{2V_b}{R}\frac{1-e^{-\pi/q}}{1-e^{-2\pi/q}} \tag{7.20}$$

and

$$A_2 = \frac{2V_b}{R}\frac{1-e^{-\pi/q}}{1-e^{-2\pi/q}}. \tag{7.21}$$

The exponential term in Equations 7.20 and 7.21 is an identity from hyperbolic trigonometry:

$$\frac{1-e^{-x}}{1-e^{-2x}} = \frac{1}{2} + \frac{1}{2}\tanh\left(\frac{x}{2}\right). \tag{7.22}$$

With the relationship of Equation 7.22, Equations 7.20 and 7.21 become

$$A_1 = -\frac{V_b}{R}\left[1+\tanh\left(\frac{\pi}{2q}\right)\right] \tag{7.23}$$

and

$$A_2 = \frac{V_b}{R}\left[1+\tanh\left(\frac{\pi}{2q}\right)\right]. \tag{7.24}$$

Substitution of Equations 7.23 and 7.24 into Equations 7.14 and 7.15, respectively, yields the expressions for the instantaneous inductive load current of the square-wave inverter:

$$i_1(\theta) = \frac{V_b}{R} - \frac{V_b}{R}\left[1+\tanh\left(\frac{\pi}{2q}\right)\right]e^{-\theta/q}, \ 0 \le \theta \le \pi. \tag{7.25}$$

$$i_2(\theta) = \frac{V_b}{R}\left[1+\tanh\left(\frac{\pi}{2q}\right)\right]e^{-(\theta-\pi)/q} - \frac{V_b}{R}, \ \pi \le \theta \le 2\pi. \tag{7.26}$$

The instantaneous current is plotted in Figure 7.11.

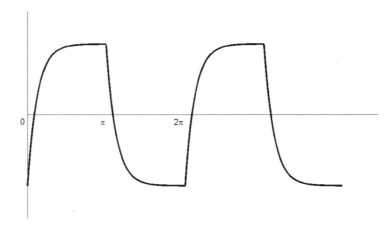

FIGURE 7.11   Instantaneous load current of the square-wave inverter.

Average Load Power

There are two ways to compute the average load power. One way is to determine the RMS value of the current expressed by Equations 7.25 and 7.26 and then use $P = I_{RMS}^2 R$, in which R is the real part of the load. The other way is to determine the power delivered by the source. If the switches are ideal, the load power and source power are equal. Since the source voltage is constant, the average power is the product of the battery voltage and the average source current. The instantaneous source current is illustrated in Figure 7.12.

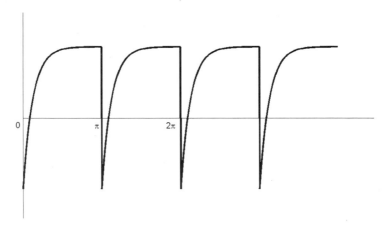

FIGURE 7.12   Source current of the square-wave inverter with inductive load.

Application of the average value integral to the waveform in Figure 7.12 results in

$$I_b = \frac{V_b}{\pi R} \int_0^\pi \left\{ 1 - \left[1 + \tanh\left(\frac{\pi}{2q} \right) \right] e^{-\theta/q} \right\} d\theta. \tag{7.27}$$

Evaluation of the integral in Equation 7.27 yields the average battery current as

$$I_b = \frac{V_b}{R} \left[1 - \frac{2q}{\pi} \tanh\left(\frac{\pi}{2q} \right) \right]. \tag{7.28}$$

Thus, the power supplied by the battery and absorbed by the load is

$$P_b = \frac{V_b^2}{R} \left[1 - \frac{2q}{\pi} \tanh\left(\frac{\pi}{2q} \right) \right]. \tag{7.29}$$

EXAMPLE 7.4

Deduce an expression for the RMS load current from Equation 7.29.

Solution

With ideal switches, the source and load power are equal:

$$P_b = I_L^2 R. \tag{7.30}$$

The solution of Equation 7.30 for the RMS load current is

$$I_L = \sqrt{\frac{P_b}{R}}. \tag{7.31}$$

Substitution of Equation 7.29 into Equation 7.31 yields the expression

$$I_L = \frac{V_b}{R} \sqrt{1 - \frac{2q}{\pi} \tanh\left(\frac{\pi}{2q} \right)}. \tag{7.32}$$

Conclusion

By the principle of energy conservation, the expression for the RMS load current is obtained directly from the source power without integration.

EXERCISE 7.4

Compute the average power absorbed in the load of a 60 Hz square-wave inverter with $V_b = 12$ VDC, $R = 1\ \Omega$, and $L = 1$ mH.

Answer

$P = 110$ W.

Efficiency

Inverters typically operate from a low-voltage, high-current source. The MOSFET is well-suited for low-voltage, high-current applications because of its low on-resistance. In addition, MOSFETs with low break-down voltages are significantly less costly than those with large breakdown voltages. Furthermore, the circuitry that drives the MOSFET H-bridge is less expensive than the drive circuitry for a BJT H-bridge.

Since two MOSFETs are always in series with the load, the total resistive component of the source impedance is

$$R = R_L + 2R_{on}, \tag{7.33}$$

in which R_L is the real part of the load impedance. The power delivered by the source is

$$P_b = I_L^2 \left(R_L + 2R_{on} \right), \tag{7.34}$$

and the efficiency is the ratio of the load power to the source power:

$$\eta = \frac{R_L}{R_L + 2R_{on}}. \tag{7.35}$$

The source power expressed by Equation 7.34 represents the sum of the load power and the power dissipated in *four* MOSFETs. The power dissipated in one MOSFET is thus

$$P_{xtr} = \frac{1}{2} I_L^2 R_{on}, \tag{7.36}$$

and the RMS MOSFET drain current is

$$I_\mathrm{D} = \frac{V_\mathrm{b}}{R} \sqrt{\frac{1}{2} - \frac{q}{\pi} \tanh\left(\frac{\pi}{2q}\right)}. \tag{7.37}$$

EXAMPLE 7.5

Compute the source power, load power, MOSFET dissipation, and efficiency of the circuit in Exercise 7.4 if the H-bridge is constructed of MOSFETs with an on-resistance of 10 mΩ.

Solution

The computations are performed by the code in Listing 7.3. The program is available on the CD-ROM as Example7_5.m in the Examples and Listings subfolder of Chapter 7.

LISTING 7.3 Solution Code for Example 7.5

```
close all, clear all, clc
Vb = 12;
RL = 1;
w = 120*pi;
L = 0.001;
Ron = 0.01;
R = RL + 2*Ron;
q = w*L/R;
x = 2*q/pi;
IL = Vb/R*sqrt(1 - x*tanh(1/x));
IL2 = IL^2;
PL = IL2*RL
Pxtr = 0.5*IL2*Ron
Ps = PL + 4*Pxtr
eta = PL/(PL + 4*Pxtr)
```

Conclusion

The resistance used in the computations must be the total resistance of the source impedance expressed by Equation 7.33. The computed results are $P_\mathrm{L} = 106$ W, $P_\mathrm{XTR} = 0.5$ W, $P_\mathrm{b} = 108$ W, and $\eta = 98\%$.

EXERCISE 7.5

Compute the RMS MOSFET drain current of the inverter in Example 7.5.

Answer

$I_D = 7.3$ A.

INVERTER SWITCHING METHODS

The switching scheme presented earlier is suitable for inverters in very low-power applications. In higher-power applications, the switching scheme is more complex to prevent circuit faults caused by the finite switching times of the power devices. A *shoot-through* fault occurs, for example, when SW1 and SW4 or SW2 and SW4 are closed simultaneously. This situation arises, for example, if SW1 and SW3 are switched on when SW2 and SW4 have not yet fully turned off.

To prevent a short circuit across the battery, the switches are actuated in a sequence that allows time for a complete transition between the on and off states. The sequence results in a *dead time,* or *blanking time,* between the time intervals during which the odd-numbered switches and the even-numbered switches are conducting. The switching sequence is presented in Table 7.1 where a "1" represents a closed switch and a "0" represents an open switch.

TABLE 7.1 Switch Sequence to Produce Blanking Time

State	1	2	3	4
SW1	1	1	0	0
SW2	0	1	1	0
SW3	1	0	0	1
SW4	0	0	1	1

With reference to Figure 7.10, the circuit operates as follows: in state 1 both SW1 and SW3 are closed and the battery voltage appears across the load. In state 2 SW3 is opened and SW2 is closed; the load is effectively short-circuited and the output voltage is zero. In state 3 SW2 and SW4 are closed and the load voltage polarity is reversed. State 4 has switch SW2 open and SW4 closed to short-circuit the load once again and output zero voltage. The load voltage waveform that results from the switching scheme in Table 7.1 is illustrated in Figure 7.13.

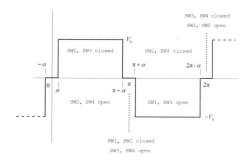

FIGURE 7.13   Load voltage with blanking time.

Load Voltage Harmonics

The simplicity of the inverter switching scheme, however, is not without consequences. The nonsinusoidal output voltage of the inverter produces current harmonics that erode load power factors and cause torque variations in AC motors. Knowledge of the harmonic content is useful in the design of more complex switching schemes and in the design of filters to remove some of the more troublesome harmonics. The voltage harmonics are readily determined with a Fourier series expansion of the waveform in Figure 7.13. Since the waveform possesses odd symmetry, the cosine coefficients are zero and the sine coefficients are determined from

$$b_n = \frac{2}{\pi} \int_{\alpha}^{\pi-\alpha} V_b \sin(n\theta) d\theta. \tag{7.38}$$

Evaluation of the integral in Equation 7.38 results in the sine coefficient expression

$$b_n = \frac{4V_b}{n\pi} \cos(n\alpha). \tag{7.39}$$

Equation 7.39 reveals that the magnitudes of the harmonics are dependent upon the dead time. It is thus possible to control the harmonic magnitudes with a variable blanking interval. Since the desired output of the inverter is a sinusoidal voltage, the magnitude of the fundamental component is of particular interest:

$$b_1 = \frac{4V_b}{\pi} \cos\alpha. \tag{7.40}$$

Output voltage regulation is thus available with a variable α. It is also possible to eliminate selected harmonics when the cosine argument of Equation 7.39 is 90°. The magnitude of the nth harmonic is zero if

$$\alpha = \frac{\pi}{2n}. \tag{7.41}$$

Voltage regulation and harmonic elimination are not independent operations, however, because the value of n must be fixed to eliminate a particular harmonic. A fixed value of n results in a fixed value of α and a fundamental component with constant amplitude. Regulation and harmonic control is possible, however, with PWM techniques.

SIMULATING INVERTERS IN MATLAB

The simulation method of the inverter is similar to that of the DC-DC converter. The inverter control switch, however, is tri-level as opposed to the binary switch introduced earlier. In terms of the control switch, the differential equation that describes the inductively loaded inverter may be expressed as

$$\frac{di}{d\theta} + \frac{i}{q} = \frac{V_b}{qR} u(\theta), \tag{7.42}$$

in which the control variable u is defined as

$$u(\theta) = \begin{cases} 0, & -\alpha < \theta < \alpha \\ 1 & \alpha < \theta < \pi - \alpha \\ 0 & \pi - \alpha < \theta < \pi + \alpha \\ -1 & \pi + \alpha < \theta < 2\pi - \alpha \end{cases}. \tag{7.43}$$

In Equation 7.43 angle α represents half the dead-time interval; 2α is the total dead time. The differential equation is best simulated when normalized with respect to the fundamental frequency of the inverter output voltage. In terms of time variable t, Equation 7.42 becomes

$$\frac{di}{dt} + \frac{R}{L} i = \frac{V_b}{L} u(t). \tag{7.44}$$

Application of the normalization process to Equation 7.44 yields

$$\frac{di}{d\tau} + \frac{R}{Lf} i = \frac{V_b}{Lf} u(\tau).$$

(7.45)

Equation 7.43 is normalized upon division of the angles by 2π:

$$u(\tau) = \begin{cases} 0, & -\delta < \tau < \delta \\ 1 & \delta < \tau < \frac{1}{2} - \delta \\ 0 & \frac{1}{2} - \delta < \tau < \frac{1}{2} + \delta \\ -1 & \frac{1}{2} + \delta < \tau < 1 - \delta \end{cases},$$

(7.46)

in which,

$$\delta = \frac{\alpha}{2\pi}.$$

(7.47)

In Equations 7.46 and 7.47, variable δ represents one-half the ratio of the dead time interval to the total period.

EXAMPLE 7.6

Simulate the inverter in Exercise 7.4 with a 30° blanking interval.

Solution

ON THE CD

The solution is provided by the programs in Listings 7.4 and 7.5 and are available on the CD-ROM as Example7_6.m in Chapter 7\Examples and Listings and inverter.m in Chapter 7\Toolbox.

LISTING 7.4   Code to Simulate the Inverter in Example 7.6

```
close all, clear all, clc
global A B d
mH = 1E-3;
T = 360;
f = 60;
L = 1*mH;
```

```
Lf = L*f;
R = 1;
A = -R/Lf;
alpha = 15;
d = alpha/T;
Vb = 12;
B = Vb/Lf;
x0 = 0;
[t, IL] = ode23('inverter', [0 4], x0);
u = 0.5*(sign(t - fix(t) - d) - sign(t - fix(t) - (0.5 - d)) - ...
sign(t - fix(t) - (0.5 + d)) + sign(t - fix(t) - (1 - d)));
subplot(2,1,1), stairs(t, u*Vb), grid
title('Load Voltage')
subplot(2,1,2), plot(t, IL), grid
title('Load Current')
xlabel('cycles')
```

LISTING 7.5   Function File for Listing 7.4

```
function dx = inverter(t, x)
global A B d
u = 0.5*(sign(t - fix(t) - d) - sign(t - fix(t) - (0.5 - d)) - ...
    sign(t - fix(t) - (0.5 + d)) + sign(t - fix(t) - (1 - d)));
dx = A*x + B*u;
```

Conclusion

Since the portions of the period are expressed in fractions, it is convenient to use units of degrees for the fundamental period and the blanking interval. The RHS of the control switch *u* must be multiplied by 0.5 because the statement composed of the *sign* functions has peak values of 2 and −2 rather than 1 and −1. The load voltage and load current are plotted in Figure 7.14.

EXERCISE 7.6

Use the MATLAB function *trapz* to compute the RMS load current of the inverter in Example 7.6 and use the RMS value to calculate the average power absorbed in the load.

Answers

$I_{RMS} = 9.9$ A and $P = 99$ W.

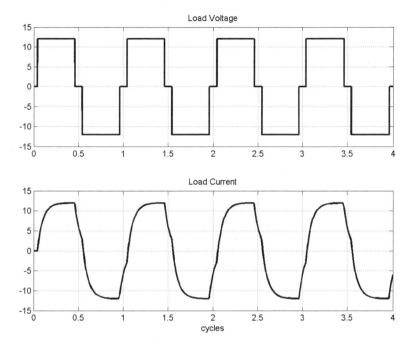

FIGURE 7.14   Load voltage and current of the inverter in Example 7.6.

Inverters with Transformers

As stated earlier, the load of an inverter circuit is typically the primary winding of a transformer—usually a step-up transformer. The schematic of a full-bridge inverter with a step-up transformer is shown in Figure 7.15. The diodes, although present in the circuit, are omitted for functional clarity.

The voltage of interest in the inverter-transformer circuit is the secondary voltage across the load. If the transformer is modeled as ideal, then the load impedance is referred to the primary winding, and the simulation of the circuit is the same as the inverter in Example 7.6. If the physical model of the transformer is used, the simulation is facilitated if the circuit is expressed in state-space form. The full-bridge inverter with the physical transformer model is illustrated in Figure 7.16.

To facilitate the analysis and simulation of the circuit, the secondary impedances and the load impedance are referred to the primary winding. The circuit with the referred impedances is shown in Figure 7.17.

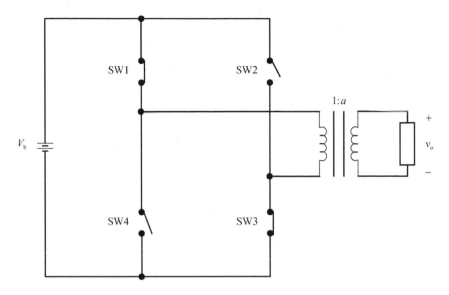

FIGURE 7.15   Inverter with step-up transformer.

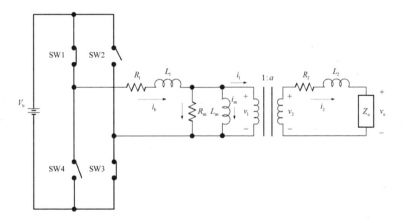

FIGURE 7.16   Full-bridge inverter with physical transformer.

With a load impedance of $\mathbf{Z}_o = R_o + j\omega L_o$, application of KCL and KVL to the circuit in Figure 7.17 yields the following set of equations:

$$u(t)V_b = i_b R_1 + L_1 \frac{di_b}{dt} + L_m \frac{di_m}{dt}. \tag{7.48}$$

$$i_b = \frac{L_m}{R_m} \frac{di_m}{dt} + i_m + i_1. \tag{7.49}$$

$$L_m \frac{di_m}{dt} = R' i_1 + L' \frac{di_1}{dt}. \tag{7.50}$$

$$R' = \frac{R_2 + R_o}{a^2}. \tag{7.51}$$

$$L' = \frac{L_2 + L_o}{a^2}. \tag{7.52}$$

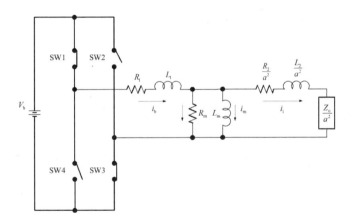

FIGURE 7.17 Inverter/transformer circuit with referred impedances.

The state variables of the circuit in Figure 7.17 are the battery current i_b, the magnetizing current i_m, and the transformer primary current i_1. The solutions of Equations 7.48 to 7.50 for the derivatives of current are as follows:

$$\frac{di_b}{dt} = -\frac{R_1}{L_1} i_b - \frac{L_m}{L_1} \frac{di_m}{dt} + \frac{V_b}{L_1} u(t). \tag{7.53}$$

$$\frac{di_{\mathrm{m}}}{dt} = \frac{R_{\mathrm{m}}}{L_{\mathrm{m}}}\left(i_{\mathrm{b}} - i_{\mathrm{m}} - i_{1}\right). \tag{7.54}$$

$$\frac{di_{1}}{dt} = \frac{L_{\mathrm{m}}}{L'}\frac{di_{\mathrm{m}}}{dt} - \frac{R'}{L'}i_{1}. \tag{7.55}$$

The state-space model must have all derivatives on the LHS of the equations. Substitution of Equation 7.54 into Equations 7.53 and 7.55 eliminates the derivative of the magnetizing current from the battery current and primary current equations:

$$\frac{di_{\mathrm{b}}}{dt} = -\frac{R_{1} + R_{\mathrm{m}}}{L_{1}}i_{\mathrm{b}} + \frac{R_{\mathrm{m}}}{L_{1}}i_{\mathrm{m}} + \frac{R_{\mathrm{m}}}{L_{1}}i_{1} + \frac{V_{\mathrm{b}}}{L_{1}}u(t). \tag{7.56}$$

$$\frac{di_{1}}{dt} = \frac{R_{\mathrm{m}}}{L'}i_{\mathrm{b}} - \frac{R_{\mathrm{m}}}{L'}i_{\mathrm{m}} - \frac{R' + R_{\mathrm{m}}}{L'}i_{1}. \tag{7.57}$$

From Equations 7.54, 7.56, and 7.57, the state-space model of the inverter/transformer circuit is formulated as

$$\frac{d}{dt}\begin{bmatrix} i_{\mathrm{b}} \\ i_{\mathrm{m}} \\ i_{1} \end{bmatrix} = \begin{bmatrix} -\dfrac{R_{1} + R_{\mathrm{m}}}{L_{1}} & \dfrac{R_{\mathrm{m}}}{L_{1}} & \dfrac{R_{\mathrm{m}}}{L_{1}} \\[2ex] \dfrac{R_{\mathrm{m}}}{L_{\mathrm{m}}} & -\dfrac{R_{\mathrm{m}}}{L_{\mathrm{m}}} & -\dfrac{R_{\mathrm{m}}}{L_{\mathrm{m}}} \\[2ex] \dfrac{R_{\mathrm{m}}}{L'} & -\dfrac{R_{\mathrm{m}}}{L'} & -\dfrac{R' + R_{\mathrm{m}}}{L'} \end{bmatrix}\begin{bmatrix} i_{\mathrm{b}} \\ i_{\mathrm{m}} \\ i_{1} \end{bmatrix} + \begin{bmatrix} \dfrac{V_{\mathrm{b}}}{L_{1}} \\[2ex] 0 \\[1ex] 0 \end{bmatrix}u(t). \tag{7.58}$$

Upon normalization, Equation 7.58 becomes

$$\frac{d}{d\tau}\begin{bmatrix} i_{\mathrm{b}} \\ i_{\mathrm{m}} \\ i_{1} \end{bmatrix} = \begin{bmatrix} -\dfrac{R_{1} + R_{\mathrm{m}}}{L_{1}f} & \dfrac{R_{\mathrm{m}}}{L_{1}f} & \dfrac{R_{\mathrm{m}}}{L_{1}f} \\[2ex] \dfrac{R_{\mathrm{m}}}{L_{\mathrm{m}}f} & -\dfrac{R_{\mathrm{m}}}{L_{\mathrm{m}}f} & -\dfrac{R_{\mathrm{m}}}{L_{\mathrm{m}}f} \\[2ex] \dfrac{R_{\mathrm{m}}}{L'f} & -\dfrac{R_{\mathrm{m}}}{L'f} & -\dfrac{R' + R_{\mathrm{m}}}{L'f} \end{bmatrix}\begin{bmatrix} i_{\mathrm{b}} \\ i_{\mathrm{m}} \\ i_{1} \end{bmatrix} + \begin{bmatrix} \dfrac{V_{\mathrm{b}}}{L_{1}f} \\[2ex] 0 \\[1ex] 0 \end{bmatrix}u(\tau). \tag{7.59}$$

The simulation of the inverter/transformer circuit is presented in the next example.

EXAMPLE 7.7

With the transformer from Example 7.3 used as a step-up transformer, simulate the inverter illustrated in Figure 7.17 with a load impedance of $Z_o = 23 + j7\ \Omega$, a battery voltage of 24 VDC, and a dead time of $20°$.

Solution

ON THE CD

The solution is provided by the code in Listing 7.6. The program is available as Example7_6.m in Chapter 7\Examples and Listings on the CD-ROM. The state variables are plotted in Figure 7.18.

LISTING 7.6   Code to Simulate the Inverter/Transformer Circuit

```
close all, clear all, clc
global A B d
Vb = 24;
d = 1/18;
f = 60;
a = 5;
uH = 1E-6;
w = 2*pi*f;
R1 = 0.02;
R2 = 0.1;
Rm = 200;
L1 = 10*uH;
L2 = 50*uH;
Lm = 0.16;
ZL = 23 + j*7;
Ro = real(ZL);
Lo = imag(ZL)/w;
Rp = (R2 + Ro)/a^2;
Lp = (L2 + Lo)/a^2;
L1f = L1*f;
Lmf = Lm*f;
Lpf = Lp*f;
a11 = -(R1 + Rm)/L1f;
a12 = Rm/L1f;
a21 = Rm/Lmf;
a31 = Rm/Lpf;
a33 = -(Rp + Rm)/Lpf;
A = [a11 a12 a12; a21 -a21 -a21; a31 -a31 a33];
B = [Vb/L1f; 0; 0];
x0 = [0; 0; 0];
```

```
tic
[t, X] = ode23s('inverter', [0 4], x0);
toc
Ib = X(:, 1); Im = X(:, 2); I1 = X(:, 3);
I2 = I1/a;
u = sign(t - fix(t) - d) - sign(t - fix(t) - (0.5 - d)) - ...
    sign(t - fix(t) - (0.5 + d)) + sign(t - fix(t) - (1 - d));
dX = A*X' + B*u';
dI1 = dX(3,:);
dI2 = dI1/a;
Vo = I2*Ro + Lo*dI2';
save data Vo t
subplot(2,2,1), plot(t, Ib), grid
title('Battery Current')
subplot(2,2,2), plot(t, Im), grid
title('Transformer Magnetizing Current')
subplot(2,2,3), plot(t, I2), grid
title('Load Current')
xlabel('cycles')
subplot(2,2,4), plot(t, Vo), grid
title('Load Voltage')
xlabel('cycles')
```

Conclusion

To use the transformer from Example 7.3 as a step-up transformer, the values of the original primary and secondary impedances have been interchanged. Also, the magnetizing inductance and resistance have been referred to what is now used as the primary winding.

The differential equations in this example are known in texts on numerical methods as *stiff equations*. Stiff differential equations are characterized by small time constants that lead to large-magnitude derivatives near the initial conditions. For example, if $x(t) = 20e^{-300,000t}$, the derivative of $x(t)$ is $\dot{x}(t) = -6,000,000e^{-300,000t}$. The magnitude of the derivative near $t = 0$ is very large. As stated previously, large derivatives cause *ode23* to reduce the time-step to maintain accuracy and result in long simulation times. Equations of this variety are best solved with a stiff differential equation solver—the type of algorithm employed in the MATLAB function *ode23s*.

Since the output voltage of the inverter is dependent upon the derivative of i_2, the state derivative is computed once again after the simulation by the statement $dX = A*X' + B*u'$. The transpose operator is required to validate the matrix products. The *save* command is used to write the output voltage and time data vectors in a file named *data.mat* for future processing. The results of the simulation are plotted in Figure 7.18.

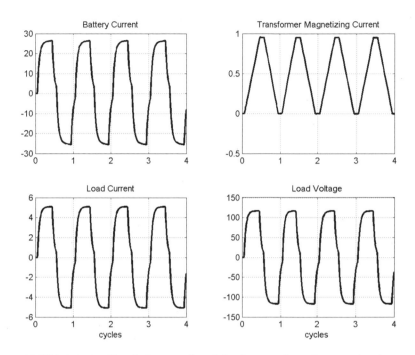

FIGURE 7.18   Simulation results of the inverter/transformer circuit.

Exercise 7.7

Use the *trapz* function to compute the output power of the inverter in Example 7.7.

Answer

$P = 400$ W.

Load Voltage Harmonics

The load voltage simulation results shown in Figure 7.18 indicate that the physical transformer acts as a filter to a small degree. The harmonic content of the output voltage is readily determined in MATLAB by use of the *fft* function, which computes the fast Fourier transform (FFT). However, before the FFT is applied to the voltage data, a bit of processing is first required. The nonconstant step-length produced by functions *ode23* and *ode23s* affects the spectrum of the computed voltage.

Most notably, because the numerical mean of the data is not zero, an artificial DC component appears in the computed spectrum. To more accurately represent the output voltage, the horizontal axis data from *ode23s* must be mapped to a constant step-length vector and the voltage data interpolated to new points that correspond to the new abscissa. The next example illustrates how to use MATLAB to interpolate data and to compute and plot the harmonic content of a vector.

EXAMPLE 7.8

Compute the FFT of the load voltage waveform of the inverter in Example 7.7 and plot the magnitude spectrum of the voltage versus frequency in hertz.

Solution

ON THE CD

The solution is provided by the program in Listing 7.7. The program is located on the CD-ROM as Example7_8.m in folder Chapter 7\Examples and Listings.

LISTING 7.7   Program to Compute the Magnitude Spectrum of the Inverter Output Voltage

```
close all, clear all, clc
load data
N = length(Vo);
ff = 60;
cycles = 4;
x = linspace(0, cycles, N);
Vo = spline(t, Vo, x);
k = 0:N - 1;
f = k*ff/cycles;
Vw = fft(Vo);
magVw = abs(Vw);
m = 1:4:49;
stem(f(m), magVw(m)/max(magVw), 'k'), grid
xlabel('frequency (Hz)')
title('Normalized Magnitude Spectrum')
```

Conclusion

The *load* statement retrieves the values of the output voltage vector **Vo** and the time vector **t** from the data file into the MATLAB workspace. Before the FFT is applied, the output voltage data is mapped to a constant step-length vector **x** by use of *spline*. This function interpolates the data in **Vo** to new points that correspond to the values contained in **x**. The result is a data vector with sample points acquired at a constant rate. The frequency axis is obtained upon multiplication of the FFT index *k* by the ratio of the fundamental frequency to the number of cycles. Each value of k thus represents a multiple of 15 Hz. Every fourth sample in the magnitude spectrum therefore represents a harmonic of the fundamental, that is, a multiple of 60 Hz. The magnitude spectrum normalized with respect to the magnitude of the fundamental component is shown in Figure 7.19.

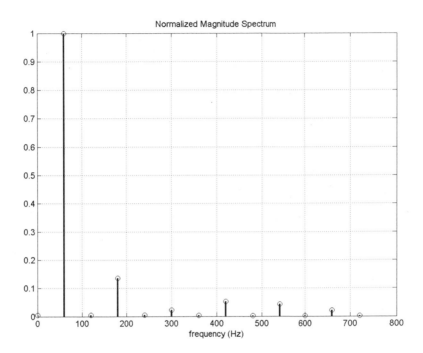

FIGURE 7.19   Magnitude spectrum of the inverter/transformer output voltage.

EXERCISE 7.8

Plot the magnitude spectrum of the output voltage from Example 7.8 without interpolation and observe the effect on the spectrum.

Filtering the Output

A capacitor is sometimes added across the secondary winding in an attempt to smooth the load voltage and render a more sinusoidal shape. Since the capacitor is in parallel with the load, the capacitor voltage is the same as the load voltage. The inverter output voltage thus becomes one of the state variables. The capacitor impedance referred to the primary winding of a transformer with a turns-ratio of $1 : a$ is expressed as

$$Z'_C = \frac{1}{j\omega a^2 C}.\qquad(7.60)$$

The equivalent capacitance on the primary side is thus

$$C' = a^2 C.\qquad(7.61)$$

With the capacitor in the circuit, the primary winding current becomes the sum of the referred capacitor current and the referred load current, as illustrated in Figure 7.20.

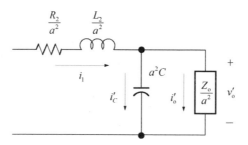

FIGURE 7.20   Inverter output with load-connected capacitor.

Application of KCL and KVL at the inverter output yields

$$i_1 = C'\frac{dv'_o}{dt} + i'_o.\qquad(7.62)$$

$$v'_o = i'_o R'_o + L'_o \frac{di'_o}{dt}.\qquad(7.63)$$

The solutions of Equations 7.62 and 7.63 for the derivatives yield the additional state equations of the inverter with an output filter capacitor:

$$\frac{dv'_o}{dt} = \frac{1}{C'}i_1 - \frac{1}{C'}i'_o.\qquad(7.64)$$

$$\frac{di'_o}{dt} = -\frac{R_o}{L_o}i'_o + \frac{1}{L'_o}v'_o.\qquad(7.65)$$

With Equations 7.64 and 7.65 included with Equation 7.58, the state-space model of the inverter/transformer circuit with load filter capacitance emerges as

$$
\frac{d}{dt}
\begin{bmatrix} i_b \\ i_m \\ i_1 \\ i'_o \\ v'_o \end{bmatrix}
=
\begin{bmatrix}
-\dfrac{R_1+R_m}{L_1} & \dfrac{R_m}{L_1} & \dfrac{R_m}{L_1} & 0 & 0 \\[2ex]
\dfrac{R_m}{L_m} & -\dfrac{R_m}{L_m} & -\dfrac{R_m}{L_m} & 0 & 0 \\[2ex]
\dfrac{R_m}{L'} & -\dfrac{R_m}{L'} & -\dfrac{R'+R_m}{L'} & 0 & 0 \\[2ex]
0 & 0 & 0 & \dfrac{R_o}{L_o} & -\dfrac{1}{L'_o} \\[2ex]
0 & 0 & \dfrac{1}{C'} & -\dfrac{1}{C'} & 0
\end{bmatrix}
\begin{bmatrix} i_b \\ i_m \\ i_1 \\ i'_o \\ v'_o \end{bmatrix}
+
\begin{bmatrix} \dfrac{V_b}{L_1} \\[2ex] 0 \\ 0 \\ 0 \\ 0 \end{bmatrix}
u(t). \qquad (7.66)
$$

Division of the RHS of Equation 7.66 by the fundamental frequency results in the normalized state-space model

$$
\frac{d}{d\tau}
\begin{bmatrix} i_b \\ i_m \\ i_1 \\ i'_o \\ v'_o \end{bmatrix}
=
\begin{bmatrix}
-\dfrac{R_1+R_m}{L_1 f} & \dfrac{R_m}{L_1 f} & \dfrac{R_m}{L_1 f} & 0 & 0 \\[2ex]
\dfrac{R_m}{L_m f} & -\dfrac{R_m}{L_m f} & -\dfrac{R_m}{L_m f} & 0 & 0 \\[2ex]
\dfrac{R_m}{L' f} & -\dfrac{R_m}{L' f} & -\dfrac{R'+R_m}{L' f} & 0 & 0 \\[2ex]
0 & 0 & 0 & -\dfrac{R_o}{L_o f} & \dfrac{1}{L'_o f} \\[2ex]
0 & 0 & \dfrac{1}{C' f} & -\dfrac{1}{C' f} & 0
\end{bmatrix}
\begin{bmatrix} i_b \\ i_m \\ i_1 \\ i'_o \\ v'_o \end{bmatrix}
+
\begin{bmatrix} \dfrac{V_b}{L_1 f} \\[2ex] 0 \\ 0 \\ 0 \\ 0 \end{bmatrix}
u(\tau). \quad (7.67)
$$

The next example presents a simulation of Equation 7.67.

EXAMPLE 7.9

Simulate the inverter from Example 7.7 with a 100-μF capacitor placed across the load impedance.

Solution

The simulation program is provided in Listing 7.8 and as Example7_9.m in sub-folder Examples and Listings of Chapter 7 on the CD-ROM.

LISTING 7.8   Program to Simulate the Inverter/Transformer Circuit with Load Filter Capacitance

```
close all, clear all, clc
global A B d
Vb = 24;
d = 1/18;
f = 60;
a = 5;
uH = 1E-6;
uF = 1E-6;
w = 2*pi*f;
R1 = 0.02;
R2 = 0.1;
Rm = 200;
L1 = 10*uH;
L2 = 50*uH;
Lm = 0.16;
ZL = 23 + j*7;
Ro = real(ZL);
Lo = imag(ZL)/w;
Lof = Lo*f;
Lop = Lo/a^2;
Lopf = Lop*f;
Rp = (R2 + Ro)/a^2;
Lp = (L2 + Lo)/a^2;
Lpf = Lp*f;
L1f = L1*f;
Lmf = Lm*f;
C = 100*uF;
Cp = a^2 * C;
Cpf = Cp*f;
a11 = -(R1 + Rm)/L1f;
a12 = Rm/L1f;
a21 = Rm/Lmf;
a31 = Rm/Lpf;
a33 = -(Rm + Rp)/Lpf;
a44 = -Ro/Lof;
a45 = 1/Lopf;
a53 = 1/Cpf;
A = [a11  a12  a12    0   0;
     a21 -a21 -a21    0   0;
     a31 -a31  a33    0   0;
       0    0    0  a44 a45;
       0    0  a53 -a53   0];
B = [Vb/L1f; 0; 0; 0; 0];
```

```
x0 = [0; 0; 0; 0; 0];
tic
[t, X] = ode23s('inverter', [0 4], x0);
toc
Ib = X(:, 1); Iop = X(:, 4); Vop = X(:,5);
subplot(2,1,1), plot(t, Ib, 'k-', t, Iop/a, 'k'), grid
title('Battery Current and Load Current')
subplot(2,1,2), plot(t, a*Vop, 'k'), grid
title('Output Voltage')
xlabel('cycles')
```

Conclusion

The battery and load currents along with the output voltage produced by the simulation are plotted in Figure 7.21.

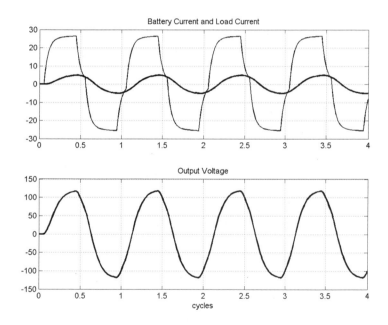

FIGURE 7.21   Simulation results of Example 7.9.

As indicated in Figure 7.21, the output voltage is much more sinusoidal with the capacitor across the load. However, the value of capacitance required to smooth the output voltage is unreasonably large. A more practical filter is considered in the next section.

EXERCISE 7.9

Repeat the simulation in Example 7.9 with a 10-μF capacitor.

Practical Inverter Filter

A more practical filter for the inverter output consists of two capacitors and an inductor, as illustrated in Figure 7.22. The filter is sometimes referred to as a "pi filter" because of its resemblance to the Greek letter.

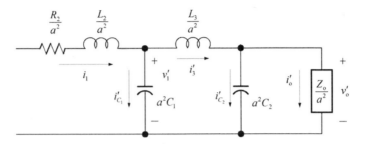

FIGURE 7.22 Inverter circuit with practical output filter.

The three additional reactive components add as many variables to the state space model. Application of KCL and KVL to the inverter circuit with the filter network results in the following set of equations:

$$L_m \frac{di_m}{dt} = R_2' i_1 + L_2' \frac{di_1}{dt} + v_1'. \tag{7.68}$$

$$i_1 = C_1' \frac{dv_1'}{dt} + i_3'. \tag{7.69}$$

$$v_1' = L_3' \frac{di_3'}{dt} + v_o'. \tag{7.70}$$

$$i_3' = C_2' \frac{dv_o'}{dt} + i_o'. \tag{7.71}$$

$$v_o' = L_o' \frac{di_o'}{dt} + i_o' R_o'. \tag{7.72}$$

The solutions of Equations 7.68 to 7.72 for the state variable derivatives are

$$\frac{di_1}{dt} = \frac{R_m}{L_2'} i_b - \frac{R_m}{L_2'} i_m - \frac{(R_m + R_2')}{L_2'} i_1 - \frac{1}{L_2'} v_1'.$$ (7.73)

$$\frac{dv_1'}{dt} = \frac{1}{C_1'} \left(i_1 - i_3' \right).$$ (7.74)

$$\frac{di_3'}{dt} = \frac{1}{L_3'} \left(v_1' - v_o' \right).$$ (7.75)

$$\frac{dv_o'}{dt} = \frac{1}{C_2'} \left(i_3' - i_o' \right).$$ (7.76)

$$\frac{di_o'}{dt} = \frac{1}{L_o'} v_o' - \frac{R_o'}{L_o'} i_o'.$$ (7.77)

Equations 7.73 to 7.76, along with Equations 7.53 and 7.54 comprise the state-space model of the inverter with the pi filter:

$$\frac{d}{dt}\begin{bmatrix} i_b \\ i_m \\ i_1 \\ v_1' \\ i_3' \\ v_o' \\ i_o' \end{bmatrix} = \begin{bmatrix} -\dfrac{R_1+R_m}{L_1} & \dfrac{R_m}{L_1} & \dfrac{R_m}{L_1} & 0 & 0 & 0 & 0 \\ \dfrac{R_m}{L_m} & -\dfrac{R_m}{L_m} & -\dfrac{R_m}{L_m} & 0 & 0 & 0 & 0 \\ \dfrac{R_m}{L_2'} & -\dfrac{R_m}{L_2'} & -\dfrac{R_m+R_2'}{L_2'} & -\dfrac{1}{L_2'} & 0 & 0 & 0 \\ 0 & 0 & \dfrac{1}{C_1'} & -\dfrac{1}{C_1'} & 0 & 0 & 0 \\ 0 & 0 & 0 & \dfrac{1}{L_3'} & 0 & -\dfrac{1}{L_3'} & 0 \\ 0 & 0 & 0 & 0 & \dfrac{1}{C_2'} & 0 & -\dfrac{1}{C_2'} \\ 0 & 0 & 0 & 0 & 0 & \dfrac{1}{L_o'} & \dfrac{R_o'}{L_o'} \end{bmatrix} \begin{bmatrix} i_b \\ i_m \\ i_1 \\ v_1' \\ i_3' \\ v_o' \\ i_o' \end{bmatrix} + \begin{bmatrix} \dfrac{V_b}{L_1} \\ 0 \\ 0 \\ 0 \\ 0 \\ 0 \\ 0 \end{bmatrix} u(t). \quad (7.78)$$

After normalization, Equation 7.78 becomes

$$
\frac{d}{d\tau}
\begin{bmatrix} i_b \\ i_m \\ i_1 \\ v_1' \\ i_3' \\ v_o' \\ i_o' \end{bmatrix}
=
\begin{bmatrix}
-\dfrac{R_1+R_m}{L_1 f} & \dfrac{R_m}{L_1 f} & \dfrac{R_m}{L_1 f} & 0 & 0 & 0 & 0 \\[2mm]
\dfrac{R_m}{L_m f} & -\dfrac{R_m}{L_m f} & -\dfrac{R_m}{L_m f} & 0 & 0 & 0 & 0 \\[2mm]
\dfrac{R_m}{L_2' f} & -\dfrac{R_m}{L_2' f} & -\dfrac{R_m+R_2'}{L_2' f} & -\dfrac{1}{L_2' f} & 0 & 0 & 0 \\[2mm]
0 & 0 & \dfrac{1}{C_1' f} & -\dfrac{1}{C_1' f} & 0 & 0 & 0 \\[2mm]
0 & 0 & 0 & \dfrac{1}{L_3' f} & 0 & -\dfrac{1}{L_3' f} & 0 \\[2mm]
0 & 0 & 0 & 0 & \dfrac{1}{C_2' f} & 0 & -\dfrac{1}{C_2' f} \\[2mm]
0 & 0 & 0 & 0 & 0 & \dfrac{1}{L_o' f} & \dfrac{R_o'}{L_o' f}
\end{bmatrix}
\begin{bmatrix} i_b \\ i_m \\ i_1 \\ v_1' \\ i_3' \\ v_o' \\ i_o' \end{bmatrix}
+
\begin{bmatrix} \dfrac{V_b}{L_1 f} \\ 0 \\ 0 \\ 0 \\ 0 \\ 0 \\ 0 \end{bmatrix}
u(\tau). \tag{7.79}
$$

A simulation of Equation 7.79 is presented in the next example.

EXAMPLE 7.10

Simulate the inverter from Example 7.7 with a pi filter with $C_1 = C_2 = 10\ \mu\text{F}$ and $L_3 = 50\ \text{mH}$.

Solution

ON THE CD

The simulation program is provided in Listing 7.9 and as Example7_10.m in Chapter 7\Examples and Listings on the CD-ROM.

LISTING 7.9 Program to Simulate the Inverter with a pi Filter

```
close all, clear all, clc
global A B d
Vb = 24;
d = 1/18;
f = 60;
a = 5;
a2 = a^2;
uH = 1E-6;
uF = 1E-6;
w = 2*pi*f;
R1 = 0.02;
R2 = 0.1;
Rm = 200;
L1 = 10*uH;
L2 = 50*uH;
Lm = 0.16;
```

```
L3 = 0.05;
C1 = 10*uF;
C2 = 10*uF;
ZL = 23 + j*7;
Ro = real(ZL);
Lo = imag(ZL)/w;
Rop = Ro/a2;
Lop = Lo/a2;
Lopf = Lop*f;
R2p = R2/a2;
L2p = L2/a2;
L1f = L1*f;
Lmf = Lm*f;
L2pf = L2p*f;
L3p = L3/a2;
L3pf = L3p*f;
C1p = a2*C1;
C2p = a2*C2;
C1pf = C1p*f;
C2pf = C2p*f;
a11 = -(R1 + Rm)/L1f;
a12 = Rm/L1f;
a21 = Rm/Lmf;
a31 = Rm/L2pf;
a33 = -(Rm + R2p)/L2pf;
a34 = -1/L2pf;
a43 = 1/C1pf;
a54 = 1/L3pf;
a65 = 1/C2pf;
a76 = 1/Lopf;
a77 = -a76*Rop;
%      Ib   Im   I1   V1p   I3p   Vop   Iop
A = [a11  a12  a12    0     0     0    0;
     a21 -a21 -a21    0     0     0    0;
     a31 -a31  a33  a34     0     0    0;
      0    0   a43    0   -a43    0    0;
      0    0    0   a54     0  -a54    0;
  0    0    0      0   a65     0  -a65;
  0    0    0      0     0   a76  a77];
B = [Vb/L1f; 0; 0; 0; 0; 0; 0];
x0 = [0; 0; 0; 0; 0; 0; 0];
tic
[t, X] = ode23s('inverter', [0 4], x0);
toc
Ib = X(:, 1); Im = X(:, 2); I1 = X(:, 3);
Iop = X(:, 4); Vop = X(:,5);
```

```
I2 = I1/a;
u = sign(t - fix(t) - d) - sign(t - fix(t) - (0.5 - d)) - ...
sign(t - fix(t) - (0.5 + d)) + sign(t - fix(t) - (1 - d));
subplot(3,2,1), plot(t, Ib), grid
title('Battery Current')
subplot(3,2,2), plot(t, Im), grid
title('Transformer Magnetizing Current')
subplot(3,2,3), plot(t, I1), grid
title('Primary Current')
xlabel('cycles')
subplot(3,2,4), plot(t, Iop), grid
title('Load Current')
xlabel('cycles')
subplot(3,1,3), plot(t, a* Vop), grid
title('Output Voltage')
xlabel('cycles')
```

Conclusion

The results of the simulation are shown in Figure 7.23.

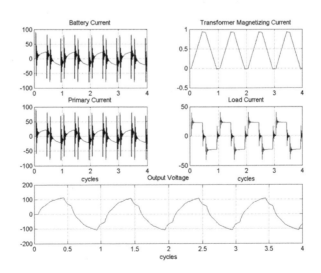

FIGURE 7.23 Results of Example 7.10 simulation.

As shown in Listing 7.9, the capacitor values are greatly reduced, but a relatively large inductance is required to achieve sufficient filtering results. In practice the values of the components used in the simulation are considered too large for economic reasons. More practical component values are obtained when PWM is incorporated into the switching scheme.

PWM INVERTER

The pi filter components in Example 7.10 are too large because of the low-frequency operation of the inverter. PWM techniques, fortunately, not only resolve the filter practicality issues but also provide output voltage control. There exist several PWM techniques, each with their own advantages and trade-offs. The method considered here is the *bipolar switching* scheme.

Bipolar Switching

Bipolar PWM makes use of a sinusoidal voltage reference voltage that is compared with a triangular *carrier* waveform. The H-bridge transistors are switched based on a comparison of the waveforms. The sinusoidal reference and triangular carrier are illustrated in Figure 7.24a with the output voltage that results shown in Figure 7.24b.

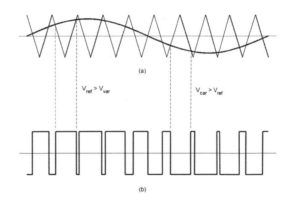

FIGURE 7.24   Bipolar switching scheme of the PWM Inverter.

In reference to Figures 7.8 and 7.24, the bipolar switching method works as follows: when the reference voltage exceeds the carrier voltage ($V_{ref} > V_{car}$), switches SW1 and SW3 are closed; SW2 and SW4 are open. The output voltage is thus $+V_b$. When the carrier exceeds the reference ($V_{ref} < V_{car}$), SW1 and SW3 are open, while SW2 and SW4 are closed. The output voltage is then $-V_b$. The period of the sinusoidal reference is set to the desired fundamental period of the output voltage. The carrier frequency is set much higher to yield smaller and more practical filter element values. A simulation of the PWM inverter is presented in the next example.

EXAMPLE 7.11

Simulate the inverter in Example 7.10 under PWM control with a carrier frequency of 12 kHz and pi filter component values $C_1 = C_2 = 1$ μF and $L_3 = 10$ mH.

Solution

Listing 7.9 is readily modified to simulate the PWM inverter with the pi filter components changed to the specified values and with the function call to *pwminver* in *ode23s* instead of the function *inverter*. The modified program is available as Example7_11.m in Chapter 7\Examples and Listings on the CD-ROM and function *pwminvr*, shown in Listing 7.10, and is available as pwminvr.m in Chapter 7\Toolbox.

ON THE CD

LISTING 7.10 Function to Simulate the PWM Inverter

```
function dx = pwminvr(t, x)
global A B d
u = 0.5*(sign(t - fix(t) - d) - sign(t - fix(t) - (0.5 - d)) - ...
    sign(t - fix(t) - (0.5 + d)) + sign(t - fix(t) - (1 - d)));
dx = A*x + B*u;
```

Conclusion

The results of the simulation are shown in Figure 7.25.

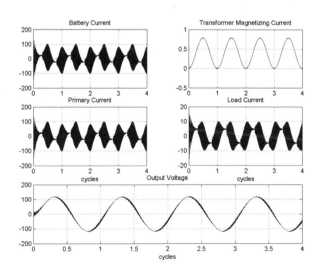

FIGURE 7.25 Results of the PWM Inverter simulation.

The simulation results demonstrate the achieved goals of the PWM control method: a sinusoidal output voltage with much smaller pi filter component values.

EXERCISE 7.11

Simulate the inverter in Example 7.11 with a carrier frequency of 25 kHz.

CHAPTER SUMMARY

The inverter is an effective circuit that produces an AC voltage in situations in which only a DC voltage source is available. For lighting and heating loads, the simplicity and economy of the square-wave inverter is ideally suited since the RMS value of the output voltage is of primary importance. If higher voltages are required than are available from the DC source, a step-up transformer is employed. In the case of an AC transformer or AC motor load, a sinusoidal output voltage is required from the inverter. In this case the more complicated but essential PWM inverter is employed to provide voltage regulation, sinusoidal wave-shaping, and practical filter component values.

THE **MATLAB** TOOLBOX

ON THE CD

The functions listed below are available on the CD-ROM in the Toolbox sub-folder of Chapter 7.

INVERTER

Function *inverter.m* computes the state derivative of the inverter state-space model with dead-time switching.

PWMINVR

Function *pwminvr.m* computes the state derivative of the inverter state-space model under PWM control.

PROBLEMS

Problem 1

Specify a transformer for a square-wave inverter that converts 12 VDC to 120 VAC with a 14.4 Ω load resistance.

Problem 2

Specify a transformer for a square-wave inverter that converts 12 VDC to 120 VAC with a 14.4 + j2 Ω load impedance that absorbs the same power as the load in the circuit of Problem 1.

Problem 3

Compute the efficiency and actual output voltage and power of the inverter in Problem 1 for a transformer that has the same physical parameters as the transformer in Example 7.3.

Problem 4

Compute the efficiency and actual output voltage and power of the inverter in Problem 2 for a transformer that has the same physical parameters as the transformer in Example 7.3.

Problem 5

Repeat Problem 3 for an inverter implemented with BJTs with a saturation voltage of 0.8 V.

Problem 6

Repeat Problem 3 for an inverter implemented with MOSFETs with an on-resistance of 10 mΩ.

Problem 7

Repeat Problem 4 for an inverter implemented with BJTs with a saturation voltage of 0.8 V.

Problem 8

Repeat Problem 4 for an inverter implemented with MOSFETs with an on-resistance of 10 mΩ.

Problem 9

Repeat Problem 1 for an inverter with a 20° blanking time.

Problem 10

Repeat Problem 2 for an inverter with a 20° blanking time.

Problem 11

Design and simulate a PWM inverter for the circuit in Problem 1.

Problem 12

Design and simulate a PWM inverter for the circuit in Problem 2.

8 | Thermal Management

In This Chapter

- Introduction
- Thermal Model of Electronic Devices
- Maximum Power Ratings of Semiconductor Devices
- Extruded Heatsinks
- Multiple Devices on a Common Heatsink
- Chapter Summary
- The MATLAB Toolbox
- Problems

INTRODUCTION

Power electronic devices typically operate at high temperatures. It is not uncommon, for example, for power devices to get hot enough to vaporize water at the surface of the device package. The heat generated by a power device must be transported to the surrounding environment at a sufficient rate so that the internal temperature of the device from does not exceed the maximum temperature specified by the manufacturer. If the maximum temperature is exceeded, the device will be destroyed. The temperature of electronic devices also affects their reliability; higher temperatures result in shorter component lifetimes. Device specifications and reliability issues thus dictate how much thermal energy must be managed to maintain the temperature of an electronic device at a safe level.

THERMAL MODEL OF ELECTRONIC DEVICES

Thermal modeling can be a very detailed and complex procedure that requires the solution of multidimensional partial differential equations. In very critical applications, the thermal computations are performed by computer programs that generate a temperature distribution profile throughout the system being modeled. Fortunately, the great majority of heat problems that arise in power electronics are solvable with very simple thermal models. The simplest model results when thermal energy is considered to flow only in one dimension under steady-state conditions.

One-Dimensional Heat Flow

The heat generated in a power electronic device occurs in the semiconductor material inside the device package. Since the semiconductor material comprises PN junctions, the internal temperature of the device is referred to as the *junction temperature*. The heat energy flows from the junction to the external device case and then flows from the case into the ambient environment. To facilitate the flow of heat, one of the power device terminals is usually attached directly to the case. For example, the anode or cathode of a rectifier may be part of the actual device case in order to promote heat flow. Thus constructed, however, the case becomes part of the electrical circuit and may require isolation from other components.

In one-dimensional heat flow the physical device is represented by *thermal resistances* through which energy flows that results from temperature differences across the resistances. The temperature differences in an electronic device occur between the junction and the case and between the case and the ambient surroundings. If the rate of energy flow is expressed in watts, a thermal resistance with units of degrees Celsius per watt experiences a temperature difference in degrees Celsius. The one-dimensional thermal model of an electronic device is illustrated in Figure 8.1.

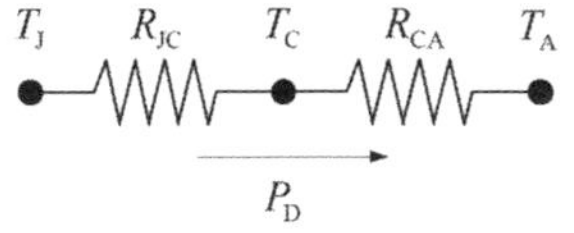

FIGURE 8.1 Thermal model of an electronic device.

In Figure 8.1 T_J, T_c, and T_A are the junction, case, and ambient temperatures, respectively. Resistance R_{JC} is the thermal resistance between the semiconductor junction and the case of the device. Resistance R_{CA} is the thermal resistance between the case and the ambient environment. The rate of thermal energy flow is P_D, the power dissipated in the device.

As indicated by the schematic in Figure 8.1, the one-dimensional heat-flow problem is analogous to an electric circuit. Temperature is analogous to voltage, power to current, and thermal resistance to electrical resistance. It is thus possible to apply KCL and KVL to solve the thermal problem. Thus applied, the circuit laws yield the junction temperature equation

$$T_J = P_D \left(R_{JC} + R_{CA} \right) + T_A. \tag{8.1}$$

Device specifications include the maximum junction temperature and the junction-to-case thermal resistance. The case-to-ambient thermal resistance is *not* specified, however. Instead, the total thermal resistance from junction-to-ambient is given; the reason will be explained shortly. The following example illustrates an application of Equation 8.1 with the specified parameters of a power electronic device.

EXAMPLE 8.1

A certain power MOSFET has a maximum junction temperature specification of 150°C, a junction-to-case thermal resistance of 1.0°C/W, and a junction-to-ambient thermal resistance of 62.5°C/W. If the ambient temperature is 30°C, what is the maximum allowable power dissipation in the device?

Solution
The solution of Equation 8.1 for the dissipated power is

$$P_D = \frac{T_J - T_A}{R_{JC} + R_{CA}}. \tag{8.2}$$

Substitution of the device parameters and the ambient temperature into Equation 8.2 yields a maximum dissipated power of

$$P_D = \frac{150 - 30}{62.5} = 1.92 \text{ W}.$$

Conclusion
The thermal resistance specified by the manufacturer is the sum of the junction-to-case and case-to-ambient thermal resistances.

EXERCISE 8.1

If the power device in Example 8.1 dissipates 2.0 W in an ambient temperature of 20°C, what is the junction temperature?

Answer

$T_J = 145°C$.

The power dissipation in the device from Example 8.1 is limited by the large case-to-ambient thermal resistance: $62.5 - 1.0 = 61.5°C/W$. Fortunately, R_{CA} can be significantly reduced with the addition of a *heatsink* to the thermal model. A heatsink is a body of material with a low thermal resistance to which the power device is attached. The material is typically aluminum, and the body is composed of fins between which air circulation occurs to remove the heat. A common heat sink profile is illustrated in Figure 8.2.

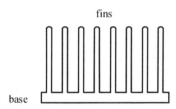

FIGURE 8.2 Heatsink profile.

The base of the heatsink provides a mounting surface for the power component as well as material depth for drilling and tapping for a machine screw with which to attach the device.

If the power device is mounted directly to the heatsink, the thermal resistance of the heatsink, R_{SA}, is placed in parallel with the case-to-ambient thermal resistance. Often, however, it is required that the power device be electrically isolated from all other components. In this case a thin thermally conductive electrical insulator is placed between the device and the heatsink. The insulator has a *case-to-sink* thermal resistance of R_{CS} and is typically about 0.5°C/W. The thermal model of an electronic device attached with a thermally conductive insulator to a heatsink is illustrated in Figure 8.3.

A common material used in thermally conductive insulators is the mineral mica. A thin layer of mica in the general shape of the device case is coated with *thermally conductive grease* to minimized the thermal resistance. *Greaseless* applications make use of a synthetic rubber insulator that is internally reinforced with a layer of

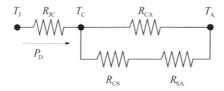

FIGURE 8.3 Thermal model of an electronic device attached to a heatsink.

fiberglass. Another type of insulator is *hard-anodized* aluminum, which can isolate voltages up to 600 V.

The thermal resistance of the insulator and the heatsink is less than the case-to-ambient thermal resistance of the electronic device—so much so, that R_{CA} is neglected in thermal calculations. As in the case of electrical circuits, if a second resistance is placed in parallel with a first resistance that has 1/10 the resistance of the first, the combined resistance is nearly equal to the smaller resistance. The same result occurs for thermal resistances connected in parallel. The case-to-sink and sink-to-ambient thermal resistances are typically much smaller than the case-to-ambient thermal resistance of an electronic device. When the sum of R_{CS} and R_{SA} is placed in parallel with R_{CA}, the case-to-ambient thermal resistance becomes insignificant. For this reason, R_{CA} is not specified by component manufactures. If the power dissipation is so small that no heatsink is needed, the total resistance from junction-to-ambient (R_{JA}) determines the junction temperature; if a heatsink is required, R_{CA} is neglected. In either case R_{CA} is an unnecessary parameter. The one-dimensional heat-flow model of the power device, insulator, and heatsink without R_{CA} is illustrated in Figure 8.4.

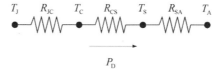

FIGURE 8.4 Thermal model of a power device with insulator and heatsink.

In Figure 8.4 T_s is the surface temperature of the heatsink. Application of KVL to the thermal model in Figure 8.4 yields the junction temperature expression as

$$T_J = P_D \left(R_{JC} + R_{CS} + R_{SA} \right) + T_A. \tag{8.3}$$

The solution of Equation 8.3 for R_{SA} is the thermal resistance of a heatsink required for a specified T_J, T_A, R_{JC}, and R_{CS}:

$$R_{SA} = \frac{T_J - T_A}{P_D} - R_{JC} - R_{CS}. \tag{8.4}$$

The next example and exercise illustrates an application of Equations 8.3 and 8.4.

EXAMPLE 8.2 .

Compute the maximum power dissipation of the device in Example 8.1 if it is attached to a heatsink with $R_{SA} = 2.5°C/W$ and an insulator with $R_{CS} = 0.5°C/W$.

Solution

The solution of Equation 8.3 for the dissipated power is

$$P_D = \frac{T_J - T_A}{R_{JC} + R_{CS} + R_{SA}}. \tag{8.5}$$

Substitution of the respective temperature and thermal resistance parameters into Equation 8.5 yields a maximum dissipated power of

$$P_D = \frac{150 - 30}{1.0 + 0.5 + 2.5} = 30 \text{ W}.$$

Conclusion

A heatsink *greatly* increases the thermal energy capacity of a power device.

EXERCISE 8.2

If the device in Example 8.2 must dissipate 40 W without an increase in junction temperature, what must be the thermal resistance of the heatsink?

Answer

$R_{SA} = 1.5°C/W$.

MAXIMUM POWER RATINGS OF SEMICONDUCTOR DEVICES

The thermal resistance of a heatsink is inversely proportional to its surface area. Hence, low-thermal-resistance heatsinks have many fins, each with large surface area. Theoretically, as the surface area approaches infinity, the thermal resistance approaches zero. The maximum power rating of an electronic device is specified by such an *infinite heatsink*. With R_{CS} and R_{SA} set to zero in Equation 8.5, the maximum power dissipation of a device is

$$P_{D,MAX} = \frac{T_J - T_A}{R_{JC}}. \qquad (8.6)$$

Since R_{CS} and R_{SA} are zero, the case temperature of the device and the ambient temperature are the same. The case temperature at maximum power is typically specified at 25°C. Although the maximum power is determined under physically impossible conditions, the specification provides an absolute limit in the choice of a power device without the need for thermal calculations. For example, if a device must dissipate 125 W in an application, then a device specified at a maximum power dissipation of 125 W *cannot* be considered for the design.

EXAMPLE 8.3

Determine the maximum power rating of the device in Example 8.1 for a case temperature of 25°C.

Solution

Directly from Equation 8.6, the maximum power is

$$P_{D,MAX} = \frac{150 - 25}{1} = 125 \text{ W}.$$

Conclusion

The device in Example 8.1 can only be used in applications that demand less than 125 W of power dissipation in the device.

EXERCISE 8.3

Determine the allowable power dissipation of the device in Example 8.1 if the device is mounted directly to a heatsink and the case temperature is 100°C.

Answer

$P_D = 50$ W.

EXTRUDED HEATSINKS

Heatsinks with low thermal resistance are made from aluminum stock that is forced under heat and pressure through a mold. This *extrusion* process typically produces 12-foot and longer lengths of heatsink profiles that are cut to smaller lengths. Figure 8.2 is indicative of the profile of a straight-fin extrusion.

The thermal resistance of the extrusion is inversely proportional to its length; the longer the extrusion, the lower the thermal resistance. The inverse relationship between length and thermal resistance, however, is not linear; doubling the extrusion length, for example, does not divide the thermal resistance in two. The thermal resistance of the extruded heatsink is typically specified under *natural convection* for a 3-inch length with the fins oriented vertically. A typical normalized relationship between extrusion length and thermal resistance is plotted in Figure 8.5.

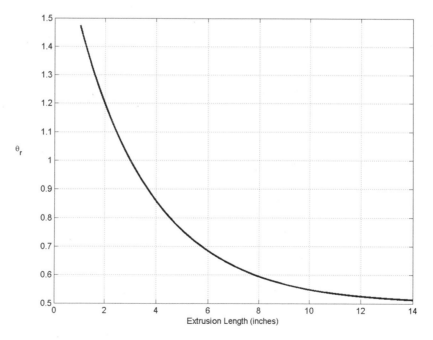

FIGURE 8.5   Normalized thermal resistance versus extrusion length.

The horizontal axis of the graph in Figure 8.5 is the length of the extrusion in inches. The vertical axis is the θ-ratio—the ratio of the 3-inch-length thermal resistance, $R_{SA,3}$ to the required thermal resistance of the application:

$$\theta_r = \frac{R_{SA}}{R_{SA,3}}. \tag{8.7}$$

As indicated in Figure 8.5, if the θ-ratio is unity, a 3-inch extrusion will meet the specification. If the ratio is greater than unity, an extrusion length less than 3 inches is acceptable. If θ_r is less than unity, an extrusion length greater than 3 inches is required. The next example illustrates the use of Figure 8.5 to determine the required extrusion length for a particular application.

EXAMPLE 8.4

The power device in Example 8.1 is used in an application in which 50 W must be dissipated in an ambient temperature of 40°C. A heatsink with an $R_{SA,3}$ of 4.5°C/W has been chosen for its width and height compatibility with the application. Determine the required length of the extrusion.

Solution

From Equation 8.4, the required thermal resistance is

$$R_{SA} = \frac{150-40}{25} - 1 - 0.5 = 2.9°C/W.$$

Equation 8.7 dictates that the θ-ratio is 2.9/4.5 $\approx$ 0.64. With Figure 8.5, the required extrusion length is determined to be approximately 7 inches.

Conclusion

The required thermal resistance R_{SA} is substantially less than $R_{SA,3}$ and results in a relatively long extrusion length.

EXERCISE 8.4

Repeat Example 8.4 with $R_{SA,3}$ = 3.4°C/W.

Answer

L $\approx$ 4 inches.

MULTIPLE DEVICES ON A COMMON HEATSINK

Product packaging and space issues often require multiple devices to be attached to the same heatsink. Each device is an additional heat source that raises the temperature of the heatsink because the dissipated power of each device flows through the same heatsink thermal resistance. The thermal model of two devices on the same heatsink is illustrated in Figure 8.6.

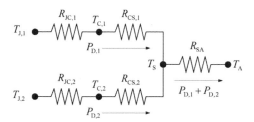

FIGURE 8.6 Thermal model of two devices on the same heatsink.

As shown in Figure 8.6, the sum of the dissipated powers must flow through the heatsink to the ambient environment. The temperature equations for the devices in Figure 8.6 are

$$T_{J,1} = P_{D,1}\left(R_{JC,1} + R_{CS,1}\right) + \left(P_{D,1} + P_{D,2}\right)R_{SA} + T_A \tag{8.8}$$

and

$$T_{J,2} = P_{D,2}\left(R_{JC,2} + R_{CS,2}\right) + \left(P_{D,1} + P_{D,2}\right)R_{SA} + T_A. \tag{8.9}$$

The solutions of Equations 8.8 and 8.9 for the heatsink thermal resistance are

$$R_{SA} = \frac{T_{J,1} - T_A - P_{D,1}\left(R_{JC,1} + R_{CS,1}\right)}{P_{D,1} + P_{D,2}} \tag{8.10}$$

and

$$R_{SA} = \frac{T_{J,2} - T_A - P_{D,2}\left(R_{JC,2} + R_{CS,2}\right)}{P_{D,1} + P_{D,2}}. \tag{8.11}$$

If the two devices are the same, with the same R_{CS} and power dissipations, Equations 8.10 and 8.11 simplify to

$$R_{SA} = \frac{1}{2}\left[\frac{T_J - T_A}{P_D} - R_{JC} - R_{CS}\right]. \tag{8.12}$$

Equation 8.12 reveals that two identical devices require a heatsink with half the thermal resistance required for one device alone. If the devices and dissipations are not identical, the required R_{SA} is the smallest value computed from Equations 8.10 and 8.11. Usually, the thermal resistance is affected most by the device with the lowest maximum junction temperature specification or by the device with the largest power dissipation. The following examples and exercises illustrate how to compute the R_{SA} of a common heatsink for two power devices.

EXAMPLE 8.5

Specify a heatsink for two identical power devices with a maximum junction temperature of 150°C to operate in an ambient temperature of 50°C. The power dissipation in each device is 30 W and the junction-to-case thermal resistance is 1.5°C/W. The thermal resistance of the insulators is 0.5°C/W.

Solution

Directly from Equation 8.12, the required thermal resistance is

$$R_{SA} = \frac{1}{2}\left[\frac{150 - 50}{20} - 1.5 - 0.5\right] = 1.5°C/W.$$

Conclusion

One device alone requires an R_{SA} of 3.0°C/W.

EXERCISE 8.5

Repeat Example 8.5 with three identical devices that dissipate 10 W each.

Answer

$R_{SA} = 2.67°C/W.$

EXAMPLE 8.6

Specify a common heatsink for two devices in an ambient temperature of 50°C. The thermal data are as follows: $P_{D,1} = 20$ W, $T_{J,1} = 150°C$, $R_{JC,1} = 1.5°C/W$, $R_{CS,1} = 0.5$ °C/W, $P_{D,2} = 30$ W, $T_{J,2} = 175°C$, $R_{JC,2} = 1.2°C/W$, and $R_{CS,2} = 0.5°C/W$.

Solution

ON THE CD

The computations are performed by the MATLAB code in Listing 8.1. The program is available as Example8_6.m in the Examples and Listings subfolder of Chapter 8.

LISTING 8.1   MATLAB Code to Specify a Common Heatsink for Two Devices

```
close all, clear all, clc
TA = 50;
PD1 = 20;
PD2 = 30;
TJ1 = 150;
TJ2 = 175;
RJC1 = 1.5;
RJC2 = 1.2;
RCS = 0.5;
RSA1 = (TJ1 - TA - PD1*(RJC1 + RCS))/(PD1 + PD2)
RSA2 = (TJ2 - TA - PD2*(RJC2 + RCS))/(PD1 + PD2)
```

Conclusion

The thermal resistances returned by the program are $R_{SA,1} = 1.2°C/W$ and $R_{SA,2} = 1.5°C/W$. Therefore, a heatsink with a thermal resistance of 1.2°C/W or lower must be used. If the larger thermal resistance is used, the junction temperature of the first device will exceed 150°C.

EXERCISE 8.6

Repeat Example 8.6 with a maximum junction temperature of (a) $T_J = 175°C$ and (b) $T_J = 150°C$ for both devices.

Answers

a. $R_{SA} = 1.5°C/W$
b. $R_{SA} = 1.0°C/W$.

Equations 8.8 to 8.12 are readily modified for N devices with a common heatsink. The junction temperature of the nth device is

$$T_{J,n} = P_{D,n}\left(R_{JC,n} + R_{CS,n}\right) + R_{SA}\sum_{n=1}^{N} P_{D,n} + T_A, \qquad (8.13)$$

and the heatsink thermal resistance based on the parameters of the nth device is

$$R_{SA} = \frac{T_{J,n} - T_A - P_{D,n}\left(R_{JC,n} + R_{CS,n}\right)}{\sum\limits_{n=1}^{N} P_{D,n}}. \tag{8.14}$$

Equation 8.14 applies to Example 8.5 for the case of $n = 2$ and to Exercise 8.5 for $n = 3$. If all thermal properties and power dissipations are equal, the required thermal resistance is

$$R_{SA} = \frac{1}{n}\left[\frac{T_J - T_A}{P_D} - R_{JC} - R_{CS}\right]. \tag{8.15}$$

Equation 8.15 is convenient in applications in which devices are connected in parallel to equally share a large load current.

EXAMPLE 8.7

Specify a heatsink extrusion length for four devices to function in an ambient temperature of 50°C. Because of reliability issues, no junction temperature is to exceed 120°C. The thermal data are as follows: $R_{SA,3} = 1.2°C/W$, $P_{D,1} = 5$ W, $R_{JC,1} = 1.8°C/W$, $P_{D,2} = 10$ W, $R_{JC,2} = 1.5°C/W$, $P_{D,3} = 15$ W, $R_{JC,3} = 1.2°C/W$, $P_{D,4} = 20$ W, and $R_{JC,4} = 1.5°C/W$.

Solution

The case-to-sink thermal resistance is not specified in the example. Typically, R_{CS} is assumed to be 0.5°C/W unless otherwise specified. The solution is provided by the program in Listing 8.2 and as Example8_7.m in Chapter 8\Examples and Listings on the CD-ROM.

ON THE CD

LISTING 8.2   Program to Compute a Common R_{SA} for Four Devices

```
close all, clear all, clc
TA = 50;
PD = [5 10 15 20];
TJ = 120;
RJC = [1.8 1.5 1.2 1.0];
RCS = 0.5;
Psum = sum(PD);
```

```
RSA1 = (TJ - TA - PD(1)*(RJC(1) + RCS))/Psum
RSA2 = (TJ - TA - PD(2)*(RJC(2) + RCS))/Psum
RSA3 = (TJ - TA - PD(3)*(RJC(3) + RCS))/Psum
RSA4 = (TJ - TA - PD(4)*(RJC(4) + RCS))/Psum
```

Conclusion

With the thermal data placed in vectors, the MATLAB code becomes very compact. The lowest thermal resistance computed by the program is 0.8°C/W, which arises from the device that must dissipate 20 W. The θ-ratio is 0.67, which, from Figure 8.5, translates to an extrusion length of approximately 6 inches.

EXERCISE 8.7

Compute the junction temperatures of each device in Example 8.7 with $R_{SA} = 0.8°C/W$.

Answers

$T_{J,1} = 102°C$
$T_{J,2} = 110°C$
$T_{J,3} = 116°C$
$T_{J,4} = 120°C$.

CHAPTER SUMMARY

The thermal calculations presented in this chapter are associated with natural convection. If lower thermal resistances are required, then *forced convection* offers a solution in which fans are used to draw air between the heatsink fins. Generally, lower thermal resistances result as the volume flow-rate of air increases. Forced convection thermal resistances are highly dependent upon the volume flow-rate; the manufacturer's specifications for a particular heatsink must be consulted for acceptable results.

Liquid cooling is also an option in thermal management. With liquid cooling, power devices are mounted on a "cold plate" through which copper tubing is routed. Coolant is pumped through the tubes, and the heat is transferred by convection to a distant radiator. Although costly, liquid cooling provides a thermal management solution in space-restricted applications.

THE **MATLAB** TOOLBOX

The functions listed below are contained in the Toolbox subfolder of Chapter 8 on the CD-ROM.

RSA

Function *rsa.m* computes the thermal resistance of a heatsink from the junction and ambient temperatures, the power dissipation of the device, and the device junction-to-case thermal resistance. The case-to-sink thermal resistance used in the calculation is 0.5 °C/W.

Syntax

```
Rsa = rsa(Tj, Ta, Pd, Rjc)
```

RNSA

Function *rnsa.m* computes the lowest thermal resistance required for N devices mounted on the same heatsink in a common ambient temperature.

```
Rsa = rnsa(TJ, Ta, PD, RJC)
```

Function arguments **TJ**, **PD**, and **RJC** are vectors that contain, in order of correspondence, the respective junction temperatures, power dissipations, and junction-to-case thermal resistances of the devices. The case-to-sink thermal resistance used in the calculation is 0.5°C/W.

PROBLEMS

Problem 1

A power device with $R_{JC} = 1.2°C$ /W dissipates 30 W in an ambient temperature of 50°C. Specify a heatsink for the device for a maximum junction temperature of 125°C.

Problem 2

A power device with $R_{JC} = 1.8°C$ /W dissipates 20 W in an ambient temperature of 50°C. Specify a heatsink for the device for a maximum junction temperature of 110°C.

Problem 3

Specify a common heatsink for the devices in Problems 1 and 2. Specify an extrusion length if $R_{SA,3} = 4.0°C/W$.

Problem 4

Specify individual heatsinks for the BJT and diode of the buck converter of Problem 5, Chapter 4. The ambient temperature is 40°C. Junction temperatures must not exceed 135°C. The junction-to-case thermal resistances are $R_{JC} = 1.5°C/W$ for the transistor and $R_{JC} = 2.1°C/W$ for the diode.

Problem 5

Specify a common heatsink for the BJT and diode of Problem 4.

Problem 6

Specify an extrusion length for Problem 5 if $R_{SA,3} = 3.5°C/W$.

Problem 7

Specify individual heatsinks for the boost converter of Problem 1, Chapter 5 if the circuit is implemented with the BJT and diode of Problem 4.

Problem 8

Specify a common heatsink for the BJT and diode of Problem 7.

Problem 9

Specify an extrusion length for Problem 8 if $R_{SA,3} = 3.5°C/W$.

Problem 10

Specify individual heatsinks for the buck/boost converter of Problem 5, Chapter 5 if the circuit is implemented with the BJT and diode of Problem 4.

Problem 11

Specify a common heatsink for the BJT and diode of Problem 10.

Problem 12

Specify an extrusion length for Problem 11 if $R_{SA,3} = 3.5°C/W$.

Appendix | **About the CD-ROM**

The CD-ROM included with *Fundamentals of Power Electronics with MATLAB* contains MATLAB script files (m-files) for many of the examples and exercises that require numerical computations. Also included are the function files that comprise the toolbox for each chapter.

CD FOLDERS

The folders and subfolders on the CD are organized by chapter number as follows:

- Chapter #
    - Examples and Listings
    - Figures
    - MATLAB Toolbox

SYSTEM REQUIREMENTS

- Windows 98/ME/NT/2000/XP
- Pentium Processor+
- CD-ROM Drive
- Hard Drive with 260 MB free disk space
- 128 MB RAM
- MATLAB 6.1 or MATLAB Student Version 6.1 or later versions

INSTALLATION

Copy the files from the CD-ROM to your choice of directory on the hard drive. Launch MATLAB and select *File, Set Path . . . , Add with Subfolders . . .* from the menu. Browse to and select the installed directory and select *ok, save,* and *close.*

Index

CD Enclosed

If this disk package seal is broken, the purchaser forfeits all return rights and privileges to the seller.